Anthropology for Development

Anthropology for Development: From Theory to Practice connects cross-cultural social theory with the concerns of development policy and practice. It introduces the reader to a set of key ideas from the field of anthropology of development, and shows how these insights can be applied to solve real-world development dilemmas.

This single, accessibly written volume clearly explains key concepts from anthropology and draws them into a framework to address some of the important challenges facing development policy and practice in the twenty-first century: poverty, participation, sustainability and innovation. It discusses classic critical and ethnographic texts and more recent anthropological work, using rich case studies across a range of country contexts to provide an introduction to the field not available elsewhere. The examples presented are designed to help development professionals reframe their practice with attention to social and cultural variables as well as understand why mainstream approaches to reducing poverty, raising productivity, delivering social services and grappling with environmental risks often fail.

This book will prove invaluable to undergraduate and postgraduate students who are professionals-in-training in development studies programs around the world. It will also help development professionals work effectively and inclusively across cultures, tap into previously invisible resources, and turn current development challenges into opportunities.

Robyn Eversole is Professor with the Centre for Social Impact, based at Swinburne University, Melbourne, Australia.

Anthropology for Development

From Theory to Practice

Robyn Eversole

Routledge
Taylor & Francis Group

LONDON AND NEW YORK

First published 2018
by Routledge
2 Park Square, Milton Park, Abingdon, Oxon OX14 4RN

and by Routledge
711 Third Avenue, New York, NY 10017

Routledge is an imprint of the Taylor & Francis Group, an informa business

© 2018 Robyn Eversole

British Library Cataloguing-in-Publication Data
A catalogue record for this book is available from the British Library

Library of Congress Cataloging-in-Publication Data
A catalog record for this book has been requested

ISBN: 978-1-138-93279-1 (hbk)
ISBN: 978-1-138-93280-7 (pbk)
ISBN: 978-1-315-67901-3 (ebk)

Typeset in Bembo
by Florence Production Ltd, Stoodleigh, Devon, UK

Contents

Figures

Boxes

Introduction

Development studies is a practice-oriented field – it is about making a difference. This field attracts students concerned about poverty and injustice, who want careers that can help make the world better.

Yet when we scratch the surface of what it means to actually *do* development work – whether as a high-level policy maker or a coalface community worker – we find that *development* itself is a slippery idea. Development can mean different things depending on who you are, and depending on what you believe about poverty and prosperity, justice and injustice.

For anthropologists, doing development must start from a basic question: *What are we trying to do, and for whom?*

Many developments

What does it mean when we say we are working to support the *development* of communities, localities, regions or nations?

Some people would say development means equipping them with what they need to become more productive and prosperous. Others would say it is about ensuring that people have the capabilities and space to create their own opportunities. Some definitions of development would emphasise education or health services, and others livelihoods and sustainability. Others would stress citizen voice, representation and rights.

All of these are valid approaches to *development*. All are about finding ways to change things for the better. Yet each idea proposes a very different kind of change. Some ideas about development focus on 'the economy', some on 'livelihoods', some on 'empowerment', some on 'access to services', some

1

on 'rights'. . . and the list goes on. Different people have different ideas about what kind of change is required and how to achieve it.

The Khumi people in the hills of Bangladesh have their own local idea of development which they call nebu-heina or better-life. For the Khumi, 'development' includes some aspects of non-Khumi education, lifestyles and livelihoods, but still remains firmly grounded in the values and lifestyle of Khumi communities.[1]

People have different views about development, and they organise their development work in a range of ways. First, there are *international* organisations dedicated to development. These include multilateral organisations working across multiple countries, such as the United Nations; bilateral organisations creating country-to-country linkages, such as USAID, DFID, GTZ or SIDA;[2] and international NGOs (non-governmental organisations) such as Oxfam, Save the Children or World Vision. These international development organisations focus on providing development assistance to poor places around the world.

Then, there are *national* organisations dedicated to development. These include national or federal government departments with a mandate to promote economic development, social development, education, health and so forth – as well as national NGOs, citizens' organisations and national industry organisations.

Finally, there are *local, regional, provincial, departmental* and *state* organisations and programs that are dedicated to promoting development at more localised scales. These organisations may concern themselves with development broadly, or with specific development topics – from education and jobs, to resource management and governance. They may be government or non-government organisations, and can be structured in a range of different ways: from voluntary progress committees to elected local governments.

All of these people and organisations talk about *development*, and everyone says something different. Anthropologists teach us that there are many different *developments* – that is, many different ways of understanding and acting to create social and economic change.

Development can be defined broadly as the process of social and economic change. Change may be intentional and planned, or unintentional and organic. People and places change – in small ways and bigger ways – all the time. While we might imagine that very isolated places or very traditional communities don't experience development, they do. Change happens over time. It is not necessarily planned or intentional, but it can create big shifts in how people live, work, and their overall standard of living. The results may be positive, negative – or a bit of both.

Development practice is the process of intentionally working to create positive social and economic change. When development professionals talk about *development work*, this is what they mean: intentional change-creating activities. Development practice includes the creation of policies and programs to make change, as well as the management and implementation of programs and projects. It includes multi-organisation partnership arrangements and multi-stakeholder initiatives, as well as initiatives undertaken by a single organisation. Development practice takes place at scales from the local to the global.

Development practice is built around a powerful idea: that positive change can be imagined and then systematically pursued. Doing development work involves imagining what a desirable change would look like – more jobs, better infrastructure, less poverty – and then putting in place a concrete strategy to get there from here.

It sounds deceptively simple. Because when it comes to imagining what 'positive change' would look like – and how to get there – there are many different views. There is no single vision of development and no pre-set path to get there. Rather, if you bring any group of policy makers or development workers into a room and ask them what they need to do to achieve *development*, you are almost certainly guaranteed a long and passionate conversation:

'First and foremost,' some will say, 'you have to get the economic signals right. If there are industries and jobs, and good infrastructure to support them, development will follow.'

'Not so,' say others. 'Jobs are important, but the economy is not the starting point. People are the heart of the economy and until they have the education, skills and confidence to drive entrepreneurship and innovation, the economy will go nowhere.'

'But you're on the wrong track completely,' others will say. 'How can you expect people to drive innovation when they can't even access the basics? Get the planning right, and sort out food, sanitation and health services if you want development.'

'But it's deeper than that,' interject others. 'You can roll out services, but that doesn't mean that everyone will be able to access them. Disadvantaged communities have rights, and they need to have a voice in how and where services are delivered. Until people have real voice in the decisions that affect them, there can be no development.'

'Maybe,' say others, 'but getting involved in politics takes time and resources. And at the end of the day people have to put their energy into practical things. Many communities create their own safety nets because mainstream systems don't work for them. We should be supporting these kinds of community-based alternatives.'

This kind of conversation could carry on late into the night, with each speaker proposing a different understanding of *development*, and a different set of ideas about how to achieve it.

Once upon a time, development meant simply growth, being modern, trying to look like someone else's view of success. Now, however, we recognise that people have different aspirations for themselves and their communities, localities, regions and nations. Everyone has different ideas about what positive change is, and how to achieve it. And they may or may not use the language of 'development' to talk about positive change.

In the twenty-first century, there are many developments – many visions of positive change, and many ideas about how to get there. To try to achieve some agreement, international policy bodies have worked to articulate some common ideas about what development is and what doing development aims to achieve. These efforts led to the articulation of the Millennium Development Goals (MDGs) in 2000 and then the Sustainable Development Goals (SDGs) in 2015.

The MDGs and the SDGs are big-picture goals that most people can agree on – things like reducing child mortality and ensuring that marginalised groups have a voice. Yet agreeing on some shared high-level goals isn't the same as agreeing on what the solutions actually look like.

Success and failure

Development practice can achieve important things. Policies, programs and projects can really make things better. They can make it possible for rural children to attend school, for instance, or for local businesses to earn more and employ more people. Development practice is based upon the compelling idea that positive change is possible and can be created.

At the same time, however, development is not always positive. Simply setting out to create positive change does not mean that positive change will happen. People may invest their scarce time, energy and resources in development projects and receive nothing in return. The pursuit of positive change can, in fact, leave people worse off than before.

How can efforts designed to create positive change ultimately do the opposite? A common theme is that ideas that look good on paper do not necessarily work well on the ground. There are several reasons why.

First, these ideas may not reflect what is really needed. People may want or need different things than the project or program provides. Developers may assume communities need schools, while in fact they need teachers. They may assume farmers lack skills, when in fact they lack access to markets. The development initiative may be based on someone else's ideas about what positive change is, or what they think is required to achieve it.

Another reason why development initiatives fail is that they misunderstand the context they are trying to change. The *context* is simply the physical and social setting where a development initiative plays out. Well-laid plans still miss important details about context: the type of physical environment, for instance, or the nature of social organisation in a particular place. A project plan may assume roads are present when they are absent, or a policy may assume that local leaders have authority they do not actually have. These kinds of mistakes are surprisingly common, and they mean that many ideas don't work at all – or create unwanted side effects that no one anticipated.

In Chiapas, Mexico, a development NGO offered a Mayan community a series of gender-sensitive livelihood projects. Yet the projects on offer — such as vegetable gardens and poultry-raising — bore little relation to the community's own aspirations. Further, they conflicted with local ways of working, creating tensions between men and women. Later, when community members were given an opportunity to plan their own development investment, they chose very different activities, and implemented them in very different ways.[3]

A further reason why efforts to create positive change may not work well is because they may be biased in favour of people and organisations that are already well off. Many projects and programs are structured in a way that benefits wealthier or more powerful people, who are in the best position to reap the benefits. These initiatives may actually look quite successful on paper: infrastructure has been built, or the overall productivity of an economy has been raised. Yet on closer look, it is clear that some people are worse off than before — often those who were disadvantaged to begin with.

For instance, a development initiative to supply improved seed to farmers may only benefit the wealthy farmers who can afford to buy it. Poorer farmers may find that they are further disadvantaged as their neighbours' businesses thrive. An initiative to provide training may only benefit those families who have political connections to the training organisation. If people's different access to resources and influence aren't considered in the design of change efforts, then poorer or less powerful people are less likely to benefit — *even when these initiatives are designed to combat poverty.*

Stories of failure are discouraging, but it is important to learn from them. It is disturbingly easy for development initiatives to favour the already-favoured, miss important contextual details and impose someone else's view of positive change where it does not fit.

Even the ideas we have about development can bias our change efforts in problematic ways. *Post-development critiques* have argued that 'development' is no longer a useful idea, because the idea of development is used to promote the agendas and worldviews of powerful countries and social groups. 'Developed' countries or communities define 'development' on their own terms – based on what they value – and then label others as 'underdeveloped'. This process actively creates inequalities.

For instance, former colonies are defined as needing to 'catch up' with their colonisers. Cultural minorities are urged to 'develop' to look more like majority groups. And historically disadvantaged social classes and communities are described as 'development problems' within more privileged societies. Groups of people are defined as inadequate, with the focus on their deficits rather than their strengths. They are pressured to pursue other people's ideas about what they need – and to do so on other people's terms.

It is a basic fact of development work that positive change for one person or group is not necessarily positive for someone else. People have different aspirations, and they also have different positions from which they experience the impacts of change initiatives. Even when development ideas seem to work well, they do not necessarily work well for everyone.

Many development initiatives produce losers as well as winners. For instance, new infrastructure like buildings or roads can create important benefits for some groups, but at the same time make life harder for others. Farmers and wage workers may benefit from a road to link them with markets and jobs. Yet hunters and pastoralists may find the road disrupts local environments and causes problems for their livelihoods.

Equally, a dam may lift energy production for a country as a whole, but displace villagers and wildlife to make room for water. New rules and institutions may benefit consumers, but place large compliance costs on small businesses. Or the rules may benefit private businesses, but make it harder for community-based groups to operate. Renewed neighbourhoods may offer a higher standard of living and healthy lifestyles, yet long-term residents may no longer be able to afford to live there.

These examples highlight that *development 'success' and 'failure' are always relative*. Development initiatives seek to create positive change, but success and failure ultimately depend on who you are.

Change and continuity

In the twenty-first century, *development* is no longer focused narrowly on growth and modernisation. When it comes to creating positive change, there is a recognition that social, cultural, political and environmental issues matter. Development practice is now more *pluralistic* and more *holistic* than before. Yet old ways of thinking are not easily discarded; they still influence, even in subtle ways, the ideas and actions of development organisations today.

Development practice in the twenty-first century is increasingly pluralistic. People now understand that different societies develop differently. Old divisions between 'global North' and 'global South' are so blurred as to be nearly meaningless. Emerging powers challenge old categories. Nations follow a range of development trajectories. Countries that once received development assistance are now providing development assistance to others. Rich countries, previously the 'developers', grapple with their own problems of poverty and growing inequity.

Further, the organisations involved in development are increasingly pluralistic too. Development is not just the responsibility of governments and government-to-government development assistance. In recent decades, NGOs, the private sector, and social enterprises have been taking a growing role in development work. Special interest groups, civil society organisations and Indigenous groups are more vocal. Governments recognise that they need to engage respectfully with stakeholders with different ideas about positive change. As a result, they increasingly work with a wide range of organisations and form multi-stakeholder partnerships.

Development practice is also more holistic. The MDGs and SDGs, as well as many national and local development policies, have moved beyond a narrow focus on economic prosperity and growth. Global development goals now consider other aspects of human well-being: from clean water and healthy environments to quality education and political institutions. It is now recognised that development must be *sustainable* – it must not privilege economic growth over the survival of environments and societies. Further, it is increasingly understood as *multidimensional* – not just about one thing, like income, but about people's ability to access and mobilise a wide range of physical, social, financial, political and other resources.

On first look, it seems that a lot has changed. Development practice in the twenty-first century appears to have taken on board many lessons from past failures and critiques. Different perspectives are recognised; multiple kinds of impacts are taken into account.

Yet if many things have changed, a lot has also stayed the same. When we consider how we do development work, there are still many similarities between development practice now, and development practice fifty or sixty years ago. Despite significant changes in how we understand development, the way that development initiatives are designed and delivered has actually changed very little.

For example, *participatory development* has become a popular idea in recent decades. Participatory development proposes that people should be involved in decisions that affect them. It reflects a recognition of pluralism. Yet in practice, participatory development often fails to deliver on the promise of multi-stakeholder decision making. People are asked to get involved in development initiatives, but they don't get much say in what happens. Development organisations work in certain ways; therefore participatory processes generally start with a pre-set agenda. Because development organisations have to work within their organisational priorities and timelines, it is difficult for them to create opportunities for authentic participation.

Sustainable development is another popular idea; it recognises that development initiatives must operate across economic, social and environmental domains. Sustainable development emphasises the holistic nature of positive change. Yet in practice, most development interventions still fail to understand or take into account their multiple impacts beyond the immediate goal in view. Development organisations are structured to work in silos: economic growth here, environmental protection there. They are sector-based rather than holistic. As a result, they tend to overlook the interconnections among social, environmental and economic domains in real-world contexts.

Finally, *rights-based development* is a recent idea that would seem to signal a commitment to shift power relations in favour of social groups who have not benefitted from past development initiatives. Rights-based development emphasises that all people have a right to such things as food, education, health care and clean environments. Development initiatives that take a rights-based approach recognise that disadvantaged groups have rights – including rights as citizens in their own countries. Yet despite a broad recognition of rights, development organisations continue to work in ways that tend to reinforce rather than challenge existing power relations.

Over and over, changes in theory are incompletely manifest in practice. This is not simply a matter of policy makers giving mere 'lip-service' to new ideas, with no real intent to do anything differently. There is in fact real commitment from many powerful people and organisations around the world to the ideas of participation, sustainability and rights. Yet the way that we 'do' development seems to regularly undermine the best of intentions. We are left with a disturbing circle: the more we change, the more we stay the same.

For those working, or planning to work, in the development field in the twenty-first century, it is no longer enough to understand how economies grow, or how development interventions can be managed and implemented. The important challenges facing development policy and practice in the twenty-first century require a deeper consideration: What are we trying to do, and for whom? It is here that the anthropology of development can help.

Development in social context

Development professionals grapple with issues and policies, target groups and program logics, organisations and outputs. Policies, programs and projects are the basic stuff of development work. But an anthropologist will tell you that development is about people.

Policies are made by people, working in organisations, guided by established institutions. Programs are designed and delivered by people, living and working in particular places and particular organisations. Projects are written and implemented by people, with particular aims and agendas. And different people benefit, or fail to benefit, from these initiatives. All development practice involves people. And thus, all development practice is located in a social context.

This may seem obvious, but in practice it is not. When we talk about 'development', we tend to speak in the abstract. We talk about the development of countries, wide regions or vaguely defined 'communities' – located in the 'global North' or the 'global South'. We talk about 'policy' as if it had legs and a face, the ability to move and act. And we talk about 'the economy', 'society' and the 'environment' in abstract terms, as if they existed on their own, rather than in specific settings where people walk, talk, and negotiate with each other about the kinds of futures they want.

Anthropology teaches us that development is a social and cultural process. Development practice at any level, from high-level policy to on-the-ground practice, takes place in real social contexts: in the boardroom, the rural enterprise, the mothers' club, the field visit, the international conference. There people interact, talking about the change they want, and proposing concrete strategies to achieve it. Development is often guided by big ideas and structures, such as the institutions of government or the market, the protocols of international trade or the rigidity of caste or class systems. Yet, ultimately, development is organised and negotiated by people.

Development actors

Development actors are the people who are involved in development. They act, and in their actions they influence the kind of development that happens (or does not happen). Development actors may be individuals, small groups, communities or organisations – small or large.

A village chief is a development actor; so is a planner in a local council, a company director or the head of the World Bank. Equally, an organisation like a local farmers' group is a development actor – as is a large international NGO. Their individual members are development actors too. The idea of the 'development actor' makes no assumption about role or status, about size or sector. Anyone can be a development actor.

A common assumption in development work is that there are only a few development actors. Development actors are assumed to be those people who have the expertise to make policy decisions about development, to design programs or projects, or to assist with the implementation of initiatives on the ground. These development actors include elected officials, technical consultants, project and program managers, directors and heads of development organisations and so forth. There are many people who hold these kinds of roles. But they are not the only development actors.

One of the lessons from decades of professional development work is that the 'beneficiaries' – those on the receiving end of development initiatives – are development actors in their own right. Beneficiaries don't just sit quietly and receive whatever development programs or projects come along. Rather, insofar as they can, they try to negotiate the terms of development.

Would-be beneficiaries may choose *not* to participate in outsiders' projects or programs – particularly if these look risky, costly, unwelcoming to people like them or ill-suited to the local context. Or, the intended beneficiaries may come along to meetings with the developers and bring ideas and projects of their own.

Beneficiaries may actively re-route project resources to benefit their own communities or families, or tweak project activities to generate different outcomes – outcomes that fit better with their own aspirations. Beneficiaries may participate in initiatives, but for very different reasons than those organising them assume. These people are more than just 'beneficiaries' of other people's actions. They are development actors in their own right.

Over and above the obvious people that are explicitly working on development initiatives, there are many other development actors. Development actors turn up in surprising places: among the faceless ranks of 'the poor', for instance, or local organisations that are invisible from a distance. Diverse development actors may be disguised under the broad umbrella of 'government' or 'industry': a bureaucrat smoothing the wheels, a small company trying something different. Development actors show up in unexpected places to drive or resist change.

Development actors do not necessarily align themselves with particular development initiatives. They may or may not describe themselves as being involved in *development* at all. As a result, people who are concerned with designing, delivering or evaluating development initiatives can easily overlook them. Many important development actors remain invisible to development professionals. Yet these actors matter when it comes to creating change.

Development actors are quick to demonstrate *agency* – that is, the ability to act. And they may act in completely unanticipated ways. Development workers often portray these unexpected actions as obstacles. When everything has been carefully planned from the top down, actors with different agendas can easily derail the best-laid development plans. On the other hand, participatory development projects and programs may prefer to see these actors as allies. They may try hard to bring these development actors on board to help design a better development product.

Yet development actors are not first and foremost either obstacles or allies for development professionals. They are people, individuals and organisations, who act on their own terms. Further, they act within the constraints and opportunities of the social contexts in which they find themselves. These social contexts stretch far beyond the limits of the development project or program.

Development projects and programs are always located within wider social frameworks and relationships. A farmers' association leader also has responsibilities to her family, her community and her own business. A project manager needs to look good to his employers, his funders and the people whom he is there to help.

All development actors – even those professionals who are dedicated to making development happen – have complex lives. Their actions for development intersect with other actions and actors, and with broader social networks and identities. These in turn influence how development happens – and, ultimately, who benefits.

Economics and beyond

The discipline of economics has long played a key role in development practice. Traditional development studies programs have had a strong focus on development economics. Economic models offer straightforward answers to complex questions about how to raise productivity, stimulate investment, raise employment or provide services in a cost-effective way. To this day, fluency in economics provides a strong advantage for securing work in development organisations.

Economic analysis plays an influential role in policy because it provides useful, concrete tools for decision making. Yet these tools can also be dangerous. Economic models are simplified views of reality; they embed assumptions which may not hold true in practice. Further, economic analyses tend to look at economic variables apart from their social and environmental contexts. When policy follows economics, and economic models get reality wrong, there are real consequences for people's lives and livelihoods.[4]

For decades, anthropologists have called attention to the limits of traditional economic analysis in explaining change. Anthropologists have revealed the gaps between economic analysis and on-the-ground economic and social realities. The sub-field of economic anthropology, in particular, has demonstrated the need to understand economic relationships within their social contexts. While much less influential than economics, anthropology has regularly shed light on the aspects of economies that economists typically miss.

Research by anthropologists challenges the myth that the 'economic' and the 'social' are separate domains. Rather, anthropologists show that economic action is deeply embedded in social institutions and relationships. This insight goes much further than just recognising the need to acknowledge 'social' aspects of development such as social safety nets, health and education, and so forth. Rather, anthropologists demonstrate that social relationships are at the core of what we call the economy.

Anthropologists' research across a wide range of contexts – from remote Amazonian tribes to urban slum communities and large development agencies – provides a wealth of information about how different people around the world 'do' economic activity. Their work allows us to see how economic activities look different in different places. Further, this research encourages us to step outside our own economic assumptions and begin to explore a wider range of economic ideas.

Encounters between buyers and sellers, employers and employees, bankers and borrowers, and so forth are all 'economic' relationships. Yet around the world, these relationships do not necessarily work in the way that an economist would expect. Economic decisions are regularly influenced by a range of contextual factors that aren't very 'economic' at all – things like social status, political agendas, values, beliefs, worldviews and so forth.

The work of anthropologists emphasises that *all* economic action – from the remote village to the corporate boardroom – takes place in social contexts. All is framed by socially situated beliefs and practices: what anthropologists call culture. Particular social groups may feel strongly about certain ideas – like *growth*, *efficiency* or *equity*. They may share certain beliefs about what is desirable and

11

what needs to be done to achieve this. And they may favour certain groups of people over others, so that some are privileged and others disadvantaged. These socially situated interactions determine how economic resources are allocated, and who ultimately benefits from them.

Because all economic action takes place in social contexts, it makes no sense to treat 'social development' and 'economic development' as separate domains, or to deal with them in policy silos. Economic development always has a social context. Often, by looking at the social context of economic activity, we can identify the roots of both social and economic disadvantage.

Policy in context?

One of the fundamental tensions in development practice is between the simplicity of policy categories and the complexity of real-world contexts. Policy categories – *gender, industry, education, environment* – tend to be abstract. They are big-picture ideas, useful concepts that bring together very different people, places and organisations into common categories. These categories provide a way to make complexity manageable. At the same time, they make complexity hard to see.

Policy categories are abstract, but change happens in the concrete world. Development projects, programs and policies hit the ground in particular physical and social settings. Each context is different. Importantly, ideas that work in one context don't necessarily work in another. The environment is different: landscapes, climates, ecosystems and infrastructures vary. The people are different, with different values and beliefs, and distinct ways of working and organising themselves. Economic and political resources are unevenly distributed from place to place.

Yet, from afar, these kinds of practical differences are hard to see. Policy makers have a bird's-eye view: a big-picture grasp of problems such as *poor health* or *unemployment*, and the needs of categories of people such as *refugee women* or *landless labourers*. Development policy defines problems and target groups, such as *improved health for refugee women*, or *employment for landless labourers*. Yet these simple, straightforward policy categories disguise the different realities that people face on the ground.

Policy categories such as 'refugee women' or 'landless labourers' contain a broad range of individuals, often with very different needs and aspirations. Some are young, some are old; some have children, some don't; some are highly skilled, some have only basic skills; some are members of community organisations, some are not. These people live in a wide range of local contexts where they may be supported and valued or ignored and abused, where they will have access to different kinds of resources and opportunities, and where they will face different kinds of problems and blockages.

From afar it is easy to generalise what people in particular categories need: *maternal health programs for refugee women*, for instance; or *enterprise development credit for landless labourers*. Looking closely at people in their local contexts, however, it may quickly become apparent that this is not what most people need at all.

For refugee women in some contexts there may already be adequate health services, but no opportunity to earn income. In other contexts, there may be good maternal health services, but no help in treating chronic disease. Landless labourers may indeed need credit, but they may need it for housing rather than for business startup. Or other issues may be much more serious in the local context, such as physical security, rendering other concerns secondary.

The distance between the high-level policy perspective and on-the-ground reality often leads to development failures. Most people who work in development can tell stories about policy initiatives that set out to make things better for disadvantaged groups, but failed to do so – and have even made things worse. Typically, when policy was made at a desk somewhere, certain things were simply not known: the monopoly that made market access impossible; the workloads that made project activities untenable; or the political dynamics that ensured poor families would be excluded from the start.

There are frequent gaps between desk-level plans, with neat links between goals, actions and results, and the realities on the ground, where these plans have very often gone awry. Conscious of past failures, policy makers increasingly recognise that effective development policy needs to pay attention to context. When trying to instigate change, the physical environment matters, and so does the social context in which a development initiative plays out.

At the same time, policy makers struggle to respond to the challenge of context. Those who guide development policy are generally located in head offices and tasked with 'big-picture' development challenges. It is hard for them to understand from afar what is really needed on the ground in different places. It is hard for them to anticipate how local contexts and actors might influence the outcomes of initiatives in unexpected ways, with unintended consequences. Context matters, yet contexts are not easy to see. It is here that anthropology can help.

An anthropological approach

About anthropology

Anthropology is the study of people. Over its history anthropology as a discipline has engaged with the variety of human experiences in different contexts all around the world. Anthropologists have spent a lot of time exploring and trying to understand the differences and similarities in people and how they live in different places. Development is just one sub-set of human experiences that anthropologists have studied.

Anthropologists are often particularly interested in observing the way of life of specific groups of people in specific local places. Traditionally, anthropologists spent long periods of time living in remote places – such as jungles, mountain villages or desert settlements – in order to understand how human societies functioned in different parts of the world.

Anthropologists have described how people in these very different places organise themselves, reckon kinship, raise children, mobilise labour and resources, ensure social cohesion, define and enforce acceptable behaviour, and understand and express their understandings about the nature of the universe. These were often very different than the ways things were done in the societies that the anthropologists were from.

These days anthropologists study people in a wide range of contexts: urban neighbourhoods, corporate offices, transnational communities, indeed the entire range of social settings where people can be found. Even very familiar social settings can be analysed to reveal the unspoken ideas, rules and structures that guide how things are done and why. And often, these still vary a lot from place to place.

To study people and how they live, anthropologists use a methodology called *ethnography*. Ethnography is a way of doing research; it involves under-taking in-depth and sometimes long-term immersion in social contexts in order to understand them. Doing ethnography means learning about groups of people by spending significant time with them. This may mean living and working with the people, speaking their language, participating in their daily activities and learning from them.

A key characteristic of anthropologists' work is that they seek to understand each form of human social organisation on its own terms. The global nature of anthropology as a discipline means that anthropologists do not start from any particular set of assumptions about what people do or don't do; about what is or isn't 'economic' behaviour; about what people 'naturally' value; or about how they 'always' organise themselves.

This means that a core characteristic of anthropology is its tendency to challenge mainstream assumptions about how things work. Anthropologists show that there is a lot of variety in how people do things from place to place. What is logical in one context may be illogical somewhere else. Things do not always work the same way. In some places, for instance, status comes from having lots of possessions – everyone assumes this is normal. In other places, however, status comes from giving possessions away.

Anthropologists like to challenge assumptions and turn things you thought you knew upside-down. This can sometimes be uncomfortable. Yet it is precisely this tendency to reframe 'big picture' questions that makes anthropology particularly helpful when there is a need to look at old problems in new ways.

Anthropology of development

Anthropologists have spent a lot of time studying why and how social and economic change happens. In particular contexts around the world, change may be slow or fast. It may be positive, negative or a bit of both. But inevitably change happens: a resource depletes, a road runs through, a new technology is discovered or someone changes the way things are done. And the results affect people.

The work of anthropologists can tell us a great deal about development, both planned and unplanned. Anthropologists have studied groups of people up-close in a wide range of local contexts, often over long periods of time. This has allowed them to observe social and economic change, planned and unplanned, and how it has played out in different places all around the world.

Anthropologists have had a unique perspective from which to understand development. In their work, they have intentionally crossed over language and cultural boundaries. They have often travelled to far-away places to observe change in action. In various parts of the world, they have used ethnography to intentionally immerse themselves in particular social contexts and understand the way of life of local people in depth. This has enabled them to observe change processes up-close and from the perspectives of members of very different groups.

By stepping into other people's 'worlds' – and leaving their own assumptions behind – anthropologists have been able to show what development looks like from the perspectives of different people in different places. They have revealed sometimes surprising perspectives on what development is. And they have been able to see, up-close, the effects of change processes and what they have meant for different groups of people.

Anthropologists look beyond broad categories such as 'local communities' or 'farming families' to take an up-close view of people's diverse situations and experiences. They show us that people all around the world have goals, but these goals, and the strategies they use to achieve them, vary. And they show us that development processes and outcomes can look quite different depending on who you are: different for women than for men, different for poor households than for wealthy ones, different for coalface project staff than central policy makers.

The anthropology of development provides a rich source of insights about how social and economic change works, and how development practice can work better. On the one hand, anthropologists studying development have called into question some of the key beliefs that development professionals hold about the people and communities they seek to help. On the other, their work has revealed a set of *actors, knowledges* and *institutions* that are often invisible in mainstream development practice, but which have the potential to reframe policy and practice in exciting ways.

Applying anthropology in development work

This book proposes that ideas from anthropology of development research can provide important practical guidance for meeting the challenges of development work in the twenty-first century. While many anthropologists have contributed to development work over the years, the rich anthropology of development literature is still not well known in development studies circles.

As a result, anthropology has been largely overlooked as a source of ideas and frameworks for development policy and practice. Now, however, is a timely moment to revisit anthropology and what it can offer to improve the effectiveness of development efforts.

In the twenty-first century, there is a growing imperative to rethink old ways of doing development. Mainstream frameworks and ways of working are proving woefully inadequate to tackle the challenges of contemporary development practice. This is apparent across all kinds of country contexts. From wealthy countries to poor ones, policy makers are under pressure to understand and respond to development needs in diverse contexts, and to build partnership with diverse social actors. Yet they are generally ill-equipped to do so.

Mainstream development practice draws heavily on economics and management approaches and frameworks. These are useful as far as they go; but they only go so far. These frameworks are of little help when policy makers are confronted with the need to grapple with complex social and cultural dynamics in very different settings. Here policy makers struggle.

Often policy makers simply tack 'social' ideas and language on to projects and programs – *gender, participation, community, sustainability, partnership, rights* – while business proceeds as usual. Too easily these ideas become merely decorative window dressing: curtains to dress up established ways of doing development.

By contrast, what is required in the twenty-first century is a reorientation of what development work looks like. This means rethinking what development is, who does it and whom it is for. The agenda of *transformational* development suggests the need to transform the systems that create and perpetuate poverty. The 2030 Agenda for Sustainable Development is called *Transforming our World*; it represents an ambitious agenda for systemic change that aims to tackle poverty and disadvantage in all its forms.[5]

Yet, for development work to be transformational, it must transform itself. It needs new frameworks, new ways of 'doing' development that can grapple with the dynamics of people in context. Experience has shown that it matters how development is defined, and who gets to define it. It matters how policy ideas 'hit the ground' in real landscapes. And it matters that some of the most dangerous ideas about development – ideas that limit people's opportunities and undermine their rights – are often the hardest to see.

The anthropology of development sheds light on the social contexts that are at the heart of all development work, from the community centre to the boardroom. Anthropology reveals that development is a social and cultural process, and provides a set of insights that can be used to understand these dynamics and create more effective development outcomes.

Aims of this book

This book has been written to introduce development students and professionals to a set of core ideas from the anthropology of development. It aims to show how anthropology can change how they think about and practise development, and equip them to address real-world development dilemmas in more effective ways.

Up until now the insights that anthropology can offer to development practice have largely remained out of reach for those without the time or inclination to delve into the anthropological literature, which is often written primarily for other anthropologists. This book provides an accessible introduction to the anthropology of development for those working, or planning to work, in the development field.

This book takes a practical approach. It recognises the reality of development failures, and the eminently practical need to ensure that future development work makes things better, not worse. The book focuses on ideas that are relevant to development practice. It then draws on case studies to illustrate what these ideas mean for real people in real places. Finally, it presents frameworks, tools and approaches that can be used to apply insights from the anthropology of development directly in development work.

Chapter 1 starts by introducing some key ideas from the anthropology of development. This chapter is about *theory* – ideas about how the world works. Chapter 1 introduces some key ideas and insights about how development works, distilled from the work of anthropologists studying development. Anthropologists' understandings of development actors, development knowledges and development institutions, taken together, provide a useful framework for understanding the social and cultural dynamics of development.

Chapter 2 shows how these ideas play out in practice, with real-world examples. This chapter looks at examples of social and economic change processes on the ground: both locally instigated 'grassroots' change, and formal planned development initiatives. The cases in Chapter 2 are from many different country contexts, wealthy and poor; they range from far-flung exotic locales to familiar towns and cities. All demonstrate practical insights and lessons from anthropological research about how development works, why many development initiatives fail and what is required for success.

Chapter 3 then draws together the insights from the first two chapters to show how an anthropological approach can be used in development practice to reorient and transform development work. The work of anthropologists offers powerful insights into the way development processes are embedded in contextualised social relationships. Chapter 3 provides a simple framework for organising core theoretical ideas from the anthropology of development and applying them in practice.

Chapter 4 goes on to show how this framework can be used in the context of day-to-day development work, to embed a deeper understanding of social and cultural context at every stage of a typical project or program cycle. This chapter illustrates how a range of existing tools and methods can be used to reveal hidden development resources, avoid costly errors and integrate a deep attention to social and cultural variables in day-to-day development practice.

Chapter 5 then looks deeper, to consider how an anthropological framework can be used to interrogate and reframe broader development challenges in the twenty-first century. Insights from the anthropology of development illustrate why mainstream approaches to reducing poverty, encouraging community participation and grappling with sustainability challenges often fail, and what can be done in response. This chapter suggests practical, anthropologically informed answers to the question of development effectiveness, and to the development challenges facing people and communities around the world.

Chapter 6 concludes the book by summing up the key characteristics of an anthropologically informed approach to development work and how this has the potential to stimulate significant innovation into the future. It concludes with three key take-home lessons that readers can carry forward into their future practical work.

Social and cultural understanding is, more than ever before, central to any effort to make the world better. Anthropology offers a unique perspective that is currently missing as development professionals grapple with the challenges of poverty and disadvantage in a diverse and rapidly changing world. The ideas in this book will help development professionals work effectively and inclusively with multiple stakeholders and communities, tap into previously invisible resources and turn current development challenges into opportunities.

Notes

1   Case study drawn from Uddin (2013).
2   USAID stands for the US Agency for International Development, DFID for the UK's Department for International Development, GTZ for the German Gesellschaft für Technische Zusammenarbeit and SIDA for the Swedish International Development Cooperation Agency.

3   Case study drawn from Simonelli and Earle (2003).
4   For instance Lawrence Haddad and colleagues from the Institute of Development Studies have observed that, 'In the past six to eight years or so key assumptions of economics have been found to do violence to reality, and hence, via poor policy choices, violence to human wellbeing' (Haddad *et al.* 2011, p. 6).
5   See UN (2015): https://sustainabledevelopment.un.org/post2015/transformingour world/publication.

1

Anthropology of development in theory

Development can be defined broadly as the process of social and economic change. Change may be planned or unplanned. Development professionals, however, are specifically interested in change that can be imagined, planned for and created. For development professionals 'development' is a planned change process that increases prosperity or well-being or social equity – depending on what kind of change they feel is most needed.

The Brundtland Commission's definition of development from the 1980s puts the emphasis on sustainable change: *development* meets human needs in the present without compromising future generations' ability to meet their needs.[1] In 2015, the 2030 Agenda for Sustainable Development identified three key pillars of development as 'people, planet and prosperity' with a focus on poverty eradication.[2] Every definition of development has a slightly different focus and emphasises different desired outcomes. But each refers to a process of social and economic change that is aiming to make the world better than it is now.

This chapter will explore some of the many ways that *development*, as planned social and economic change, is understood, practised and experienced around the world. To do this, it draws on the work of anthropologists who study development. This chapter does not attempt to canvass the entire range and richness of anthropology theory. Rather, it focuses in on some key insights from the anthropology of development which are relevant to development professionals, and which can directly inform development practice.

Development in context

Theories are sets of ideas about how the world works. Our ideas about how the world works – our theories – are important, because they guide our actions and ultimately inform what we can achieve. In development practice, our theories about social and economic change guide how we go about pursuing change. They shape the approaches we take, the tools we use and the kinds of results we look for.

Over the decades since development work has become a recognised area of professional practice, a number of powerful ideas have framed people's understandings about how development works. Key ideas like *underdevelopment, poverty, technology transfer* and *economic growth* have had a long-lasting impact on how many people think about what development is and how to achieve it.

Many ideas about development have their roots in Western philosophy; for instance, the idea of *progress* as something that can be defined and achieved over time. Other ideas, like *underdevelopment* and the *Third World*, have emerged from colonial and post-colonial history, in which wealthy countries defined what 'development' should look like for poorer countries. From the mid twentieth century, a growing development industry emphasised the idea that technology transfer should be used to stimulate economic growth. Many of these ideas still influence development practice today.

Old ideas persist, but at the same time new ideas about development are always coming onto the scene. Ideas like *good governance, sustainable livelihoods, inclusive growth* and *accountability* distil key insights that have emerged in practice about the processes and goals of positive change. These ideas from recent decades focus on more 'social' questions about power relations and who benefits from change processes.

Development has its own large vocabulary. There are always a range of ideas jostling for attention, and many of these ideas carry a lot of baggage. Some ideas about development become so popular that they develop a life of their own; they become 'buzzwords'. A buzzword can have 'a multitude of meanings and nuances, depending on who is using it and in what context'.[3] It is therefore possible for different people to use the same word – like *poverty* or *sustainability* – but use it to mean very different things.

Ideas, combined into theories, guide practical choices about development. Each choice to invest in productivity-raising technology, or to invest in new governance institutions, is informed by particular sets of ideas about what sort of change is likely to make things better, and what is required to make this change happen. Theory informs practice.

Practice informs theory too. In development work, policy makers have practical concerns that often form the starting point for development actions. They may be under pressure to create jobs, to provide infrastructure, or to ensure there are adequate services in hard-to-reach places. They ask for ideas – theories – that will show them how to create the kinds of change they need.

In development work, the link between theory and practice is not always clearly articulated. Development work is focused on practical action, getting things done. Yet, on closer look, most development policies, programs and projects are based on an idea or set of ideas about change: ideas about what change is desired, and what needs to happen in order to get there. This may or may not be clearly articulated, but it is called a *theory of change*. A theory of change is a set of ideas about how change will happen.

Development initiatives are often based on a straightforward theory of change. For instance, if the aim is to raise productivity in local microenterprises, then development workers might design a training program for enterprise owners. Training makes sense because of the idea that improving people's skills – their 'human capital' – raises productivity. Or, if the aim is to encourage people to adopt particular kinds of health practices, development workers may use a community-based extension program to make people aware of these practices and their benefits. In each case, the theory of change suggests that certain actions will – if done properly – lead to certain outcomes. Microenterprises will become more productive. Households will adopt healthier practices.

Nevertheless, even when there is a clear and logical theory of change, the desired change may not actually happen. Doing *this* may not, in practice, lead to *that*. Training may happen and yet microenterprise productivity levels may remain the same. Community-based extension activities may be beautifully implemented, but fail to make any difference. These are the problems that vex development work. Things that should work, often don't. And the theories that guide us often cannot tell us why.

Even when development initiatives make sense in theory, they may fail to work in practice. Doing this does not always lead to that because other factors are at play that can influence the results. The industries in question may have supply-chain problems, for instance, or an inability to access credit; training may therefore make no difference to their overall productivity. Households on the receiving end of extension programs may have any number of compelling reasons for rejecting new practices – such as lacking the necessary infrastructure or time – even when they understand them perfectly.

Anthropologists who study development recognise that every development initiative hits the ground in a particular context. The *context* is the on-the-ground setting in which development happens. Context is both physical and deeply social. It is the physical setting (natural and built), as well as the ways in which people work, live and interact with one another in that place.

Most theories of change assume that change is a straightforward technical process. They look at the connection between one variable (like training) and another variable (like productivity). Anthropologists, however, see change differently. Anthropologists recognise that every context is different, and these differences affect what happens. In each context, numerous variables are at play that will ultimately affect the outcomes of a development initiative. Further, the particular people involved will always influence the direction of change.

Anthropology of development theory thus starts from a key insight: Development cannot be created in isolation from its social context. Development is not a technical process in which doing *this* will always achieve *that*; rather, every development initiative is co-constructed and negotiated among people in particular contexts. These interactions affect what sort of change takes place, and who wins and who loses from development action. For anthropologists, development is first and foremost a social process.

Anthropologists and development

All around the world, from isolated villages to cosmopolitan cities, anthropologists have conducted research to understand the many different ways that people live. They have looked up-close at different contexts and what life is like for people living there. Anthropologists have been interested in kinship and social organisation, gender roles, division of labour, political institutions, beliefs and practices, and many other aspects of people's lives. They have not, however, always been interested in development.

For years nearly all anthropologists studied people whose ways of life were very different from their own. Many anthropologists in the nineteenth and twentieth centuries worked in places far away from their own homes, conducting fieldwork with groups of people who were largely unknown to them: the Yanomami, the Xavante, the Nuer and many others. The opportunity to travel and work with 'exotic' peoples was often a drawcard for would-be anthropologists.

Many anthropologists who travelled far afield for their studies assumed that they had discovered societies where people had lived unchanged for hundreds, even thousands, of years. Compared with the rapid pace of change in the places where the anthropologists were from, the people that anthropologists studied looked static and unchanging. They lived in remote places. They did not use modern technology. It was not uncommon for academics to call them 'stone age' or 'iron age' peoples, even in the twentieth century, because their way of life looked like a leftover from an earlier age.

Over time, however, anthropologists started to question this belief. Working with these groups of people up-close, all over the world, they began to observe evidence of social and economic change taking place. In the second half of the twentieth century, the world was changing quickly. New technologies were making travel and communications easier. It became clear that 'traditional' peoples were not static and unchanging. They did take on new ideas and practices.

Anthropologists also began to recognise that change wasn't something new. So-called 'traditional' peoples had always changed and adapted. They were not leftovers from earlier ages; rather, they were contemporaries with histories of their own. Eric Wolf's influential book *Europe and the People without History* documented the active histories of many social groups that had been previously portrayed as static, passive recipients of European colonisation. This book and others made the case that all communities and societies – including apparently isolated and traditional ones – actively navigate and instigate change.

Nevertheless, anthropologists still tended to have an ambivalent view of change. Anthropologists saw the value in traditional ways of working, and feared what would happen when distinctive local practices were replaced by modern cook-stoves, televisions, wage-labour and other global technologies and practices. They feared that the uniqueness of these 'traditional' cultures would be lost. When anthropologists saw the groups they worked with losing their languages or giving up their traditional lifestyles, they tended to view change as *not* a good thing.

This ambivalent view of change was exacerbated when *development* became an explicit focus for national governments and international organisations. From the mid twentieth century, anthropologists found themselves witnessing numerous intentional change-creating efforts in the places where they worked. Most of these development initiatives were explicitly focused on *modernisation*: they aimed to replace 'traditional' lifestyles with more modern ones based in industrial technology and cash economies. Modernisation as an idea portrayed traditional practices as backward and undesirable. Anthropologists took a sceptical view.

Anthropologists thus found themselves at the pointy end of development: disagreeing with its premises while directly observing its impacts. Anthropologists working in specific contexts were often in a position to witness rapid social and economic changes and what they meant for real people on the ground in real places – often far from the centres of decision-making power. Frequently, the people on the receiving end of development initiatives were given no choice. They were forced to change in ways that other people wanted.

Some of these changes were disastrous for the people concerned. Local languages were banned, villages displaced or destroyed, migratory groups were forced to be sedentary, access to forests or waterways was blocked, traditional livelihoods were replaced with precarious low-wage labour, and communities' ways of life were destroyed. Some development efforts were helpful, improving people's options, security or standards of living – but there were plenty of initiatives that did the opposite. And often, anthropologists were in a position to witness the impacts first-hand.

Because of these experiences, anthropologists have tended to have a very ambivalent view of development. Unlike economists and development managers, anthropologists have not started from the position that development is necessarily a good thing. If anything, anthropologists have started from a contrary position: that development may not be the solution, but rather the problem.

Anthropologists have often taken a critical view of development because their work has enabled them to observe, up-close, the effects of development for real people in real places. Through their experiences, they have been conscious of what happens when rapid change – planned or unplanned – threatens people's way of life. They have seen people lose their land and their livelihoods. They have seen communities uprooted and fragmented, and functioning social systems damaged beyond repair.

Over the years, many anthropologists have sought to mobilise these experiences observing development in context, and to use what they know to make a positive difference. Anthropologists, as both academics and practising anthropologists, have encouraged development workers to become more engaged with questions of social context. They have called attention to how development processes have disadvantaged particular groups of people and provided constructive suggestions to prevent this. And they have worked to help make development initiatives more inclusive of the needs and agendas of groups on the margins, all over the world.

Development up-close

Of all the social sciences, anthropology is the most intimate; it is interested in people and their lives seen up-close. Anthropologists' main research approach, *ethnography*, involves spending time immersed in particular social contexts: learning the language, getting to know the people and understanding intimately how things are done. Immersion in social contexts provides anthropologists with a unique position from which to understand development.

Using ethnography, anthropologists are often in a position to witness development processes and outcomes up-close. This up-close view has two dimensions. First, anthropologists are able to observe development 'on the ground'. They can see what is really there, rather than what outsiders simply assume is there. Next, using ethnography, they can understand development 'from within' particular social contexts. Anthropologists learn to step outside their own worldviews and see situations from the perspectives of the people they work with.

Anthropological research therefore can reveal what development looks like up-close, both 'on the ground' and 'from within'. As one writer on development recently observed, 'anthropology can contribute to development in ways that other disciplines struggle to replicate, most notably by providing situated or grounded analyses of local experiences.'[4]

Unlike those engaged with development from afar, in head offices or high-level fora, anthropologists have a front-row seat to observe development intimately. They can witness how it affects specific people in specific places, as well as how these people themselves understand what is happening and choose what they will do about it. Further, anthropologists can observe how the situations these people live in can in turn limit what they can do.

The anthropologist Sherry Ortner has written about the value of understanding social dynamics up-close. She observes:

The attempt to view other systems from ground level is the basis, perhaps the only basis, of anthropology's distinctive contribution to the human sciences. It is our capacity, largely developed in fieldwork, to take the perspective of [others], that allows us to learn anything at all. . . Further, it is our location 'on the ground' that puts us in a position to see people not simply as passive reactors to and enactors of some 'system', but as active agents and subjects in their own history.[5]

When anthropologists analyse development up-close, their findings can be challenging. The up-close view can call into question some of the key beliefs that development professionals hold about their work, and what they think it means for the people and communities they seek to help. The perspectives of local people, in particular, may reveal that 'transformative' projects are effectively keeping things the same, that 'significant' impacts failed to impact the real issue, or that 'community-led' efforts were, in the end, controlled by outside organisations.

While anthropologists' view of development can be challenging, it is an incredibly valuable way to understand how development works. In particular, this up-close view provides a way to challenge *stereotypes*, *silos* and *assumptions* – and replace them with real, empirical understanding of development contexts. Stereotypes, silos and assumptions are all unfortunately rampant in development work, and they sit at the root of many disappointing development results.

Stereotypes are broad-brush generalisations about people which do not reflect reality. Categories like 'villagers', 'farmers' or 'women' are frequently used in development practice as a shorthand for people with particular kinds of characteristics. These categories can easily become stereotypes, however, when they are used to imply that everyone in a category is basically the same: for instance, a stereotype that all *villagers* are poor, or that all *women* are mothers.

From a distance it is hard to see the diverse situations of villagers or women. Because of this, it becomes easy to substitute stereotypes for real understanding of people and their circumstances. Research by anthropologists counteracts stereotypes by providing an on-the-ground description of people in all of their diversity, as well as a from-within understanding of the social categories that matter. Anthropologists reveal the people behind the stereotypes (see Box 1.1).

Silos are the imaginary boundaries between different sectors or areas of work: such as employment, environment and health. Academics and development professionals tend to think about development in silos, because this is how development policies and strategies are generally organised: in departments, portfolios or programs, with names like 'economic development' or 'social services'. Further, development professionals tend to be specialists in a particular area: experts in health or employment, gender or housing. All of this contributes to a tendency to think and work in silos.

Box 1.1: Peasants: The people behind the stereotypes

The word *peasant* is not used a lot these days, but there was a time not long ago when nearly all small-farm projects in developing country contexts were about peasants. *Peasants* are smallholder farmers who rely on family labour to produce their own food and other subsistence products. The term 'peasant' began as a descriptive category about how farming in many parts of the world worked. Nevertheless, this term quickly became a magnet for all kinds of stereotypes. Peasants were described by developers as irrationally 'traditional', 'ignorant', 'risk-adverse', 'downtrodden', 'backwards', and incapable of instigating change on their own without outside development help. Yet anthropologists working with peasants on the ground in different parts of the world have questioned these stereotypes. They have shown peasants to be diverse, adaptive and often entrepreneurial resource producers and managers. Donald Attwood's (1997) work on 'The Invisible Peasant', for instance, showed how peasants in India were far from backward; they established a thriving, co-operative-based sugar-cane industry that outperformed the government's efforts at sugar industry development.

On the ground, however, silos make little sense, because different aspects of people's lives are interconnected: health is connected with housing, work with gender, transport with training opportunities, and so forth. Anthropologists' up-close view of development contexts can quickly break down silos to reveal important interconnections (see Box 1.2).

Assumptions are simply beliefs about local contexts which do not hold true in practice. For instance, outsiders may assume that social or environmental conditions in a particular project area are just like their own, when they are actually very different. Decision makers in urban contexts may assume that people in rural areas have access to reliable transport or communications services, when this is not the case. Policy makers from wealthy backgrounds may assume that less wealthy people can afford to pay a 'small' fee for a service that they cannot, in fact, afford.

Assumptions are typically grounded in the experiences and worldviews of outside developers. These are often very distant from the experiences and worldviews of those they seek to help. Anthropologists' up-close view can name and challenge assumptions about particular groups of people and the contexts where they operate. Challenging assumptions in turn can reframe how problems and solutions are understood (see Box 1.3).

Anthropologists' willingness to immerse themselves in diverse social settings has placed them in a unique position to observe development *in context*.

Box 1.2: Joining up the silos

Development work tends to focus on problems or issues, defined in policy silos. By contrast, an up-close view shows that different aspects of people's lives are intricately intertwined: families with employment, health with transport, and so forth. Development policies, programs and projects tend to target particular areas of people's lives, like health or employment, without acknowledging the linkages among them. Practically, this means that when initiatives are implemented and evaluated, many of their actual impacts can be missed. For example, Elizabeth Harrison has written of how an impact assessment of an irrigation project in Malawi found positive impacts of irrigated farming for household food security and poverty reduction, based on measures of household income and caloric intake. Yet the assessment of the project completely missed other impacts such as elite dominance of irrigation institutions, the scheme's impacts on land access, and the need to pay bribes to access water. Harrison observed that despite the narrow focus on increasing food production, 'The scheme does not exist in isolation from a broader political context'[6] which directly influenced farmers' options and the kinds of impacts they experienced.

Box 1.3: Challenging outsiders' assumptions

In development work, assumptions can make even hard data unreliable. Researchers unfamiliar with particular contexts may collect and interpret data based on unquestioned assumptions – and end up with an inaccurate view of what is really going on. In *Development Economics on Trial*, anthropologist Polly Hill (1986) used her work with cocoa producers in Ghana to highlight how the economists who were providing advice on rural development issues were analysing the local context through the lens of Western economic concepts like *subsistence farming* and *hired labour*. Further, they were using Western units of measurement to gather data from farmers. On the ground in rural Ghana, people organised their households, labour relations and landholdings very differently than outsiders assumed. They also used different measurement scales. As a result, official questionnaires made little sense to the local farmers, and important development decisions were proceeding on the basis of misleading data.

Anthropologists have been in a position to observe, in detail, what is actually happening on the ground in particular localities and organisations, and what development looks like from the perspectives of people with very different life experiences and worldviews. The ability to see development on the ground and from within has equipped anthropologists to unmask stereotypes, reveal connections and challenge assumptions – and to do this over and over, all around the world.

At a basic level, an up-close understanding of development is not hard to achieve. Anyone can spend time in the field, observing and listening. The idea of an *immersion*, in which development professionals spend time living with the people they aim to help, is an accepted method in development work (see Chapter 4). Yet, in practice, it is surprisingly rare for most development workers to seek out this kind of up-close view. Robert Chambers famously described a preference for 'rural development tourism', where professionals conduct brief and carefully scripted field visits to villages conveniently located on paved roads.[7] The visitors fail to see the majority of rural people or observe what life is really like in rural places.

Busy professionals are often under time pressures that limit their time in the field; but there is often more to it than that. Entering an unfamiliar context can be incredibly illuminating, but it can also be scary. It is often uncomfortable to enter other people's places, where other people's views can directly challenge what we believe about who we are and how the world works. Most people prefer to stay where they are comfortable, where they know what to expect – and where they are unlikely to be surprised.

Anthropologists in development work

Anthropology provides important insights about the on-the-ground contexts of development work and the perspectives of those involved in it. For decades, anthropologists have engaged with development work, both as academic researchers and as practising 'applied' anthropologists in development roles.[8] In both research and practice, they have contributed in-depth insights from their up-close approach. Nevertheless, anthropologists have seldom had an influential voice in development decision making, unlike other disciplines like economics.[9] And anthropology is still not well known in professional development circles.

There are several reasons why. First, critical voices are seldom popular; and anthropologists frequently give a critical view of what is happening in development work. Anthropologists tend to challenge dominant categories and assumptions, and to privilege the less powerful voices of people on the margins. As a result, development professionals may feel as though their work is being unfairly criticised, or that their professional knowledge is being undervalued. Accomplished professionals may struggle to understand why they need to take on board perspectives that are uncomfortably different than their own.

Next, there is a difference of aim. Anthropologists tend to want to understand what is going on in a particular social context; development professionals want to know what they can do about it. That is, development professionals are basically action-oriented and want quick answers, while anthropologists are more interested in the details and taking enough time to understand what is really going on. To development professionals, anthropologists may come across as slow and impractical when there are tight deadlines to be met.

Finally, and related to the second point, development professionals and anthropologists have a different approach to knowledge. Developers' focus on practical action for change means that they tend to view knowledge as something that must have a practical purpose. When knowledge does not appear to have an immediate practical application, it is not perceived as usable or relevant. As the anthropologist Maia Green has observed, 'The institutional space simply does not exist in development for the incorporation of anthropological knowledge.'[10] Policy makers working with anthropologists have articulated their frustration as: 'We don't want to know that it's all very complex. We want to know what to do.'[11]

As a result of these differences, it has not always been easy for anthropologists to contribute to practical development work. Some anthropologists have therefore kept their distance. Others, however, have made an effort, despite the obstacles, to contribute insights from anthropology to inform development work. Anthropologists have engaged with development both directly and indirectly: in professional development roles, and as academic researchers contributing their insights. Anthropologists have engaged with development in three broad ways.

First, many anthropologists study social and economic change, well beyond the scope of planned development initiatives. These are anthropologists of development 'with a small d' – anthropologists who study change over time.[12] Their work helps us understand what drives social and economic change. As researchers, they seek to understand why and how development processes occur and what they mean for different groups of people. They document how people around the world organise themselves to create, resist or adapt to change, and how structural, historical and political factors from beyond the local area interplay with local development action. Anthropology of development has often provided key background information to inform the planning of intentional development initiatives.

Next, some anthropologists study development practice. The second group can be thought of as anthropologists of Development 'with a big D': those who study the intentional change-making efforts of agencies, governments, NGOs and professionals.[13] These anthropologists analyse development initiatives so as to critique, inform and/or improve them. Often, they seek to understand the *ideas* and *practices* of development; that is, how development organisations and those who work in them think, speak, and organise themselves to create change 'for' or 'with' others.[14] Most anthropology of Development has focused on international development assistance; however, there is a growing tradition of anthropologists analysing other development contexts such as local, regional and community development practice.[15]

Finally, some anthropologists apply their anthropological knowledge and approaches in practical development work. They are sometimes referred to as Development anthropologists, or applied anthropologists of Development.[16] These anthropologists may work as applied researchers, full-time or part-time consultants, professional staff in development organisations, advisors in government departments and so forth. Development anthropologists are actively involved in designing, delivering and/or evaluating development initiatives. They apply anthropological approaches to help organisations and communities improve how they 'do' development.

Each of these three categories represents a distinct way that anthropologists engage with development. Many do so in more than one way. Anthropologists may study both social change and planned Development; or they may combine their academic research with practical development work. When it comes to engaging with development, anthropologists often disagree with each other about where their priorities should lie. Should they focus on studying and understanding change processes broadly, or analysing specific change-creating efforts? Should they maintain a critical distance to inform better policy, or should they get directly involved in practical development work? There are good reasons for doing all of these things; each contributes in different ways.

This book draws upon the contributions of all three approaches. All provide important insights for development practice. Despite their differences, some common insights are found in the work of all anthropologists who engage with development. All are interested in people, and all recognise that development affects, and is affected by, people. All are interested in what development looks like 'on the ground' and 'from within' contexts that may be quite different to our own. In the end, the work of anthropologists shows us that development is always about real people in real places. It is not a predictable technical process, but a negotiated social one.

Development actors: The people in the process

Anthropologists are interested in people: who they are, what they do, how they understand the world – even how they talk about ideas like 'development'. Indeed, there is an important genre within anthropology of development that is primarily concerned with development discourse – the language people use to talk about development.[17] For anthropologists, development cannot be separated from its social context. People – their ways of thinking, organising and relating to one another – are central to understanding how and why societies and economies change.

When anthropologists analyse development, they often focus on the people that are involved in a particular change process. These people are referred to as *development actors*. Development actors may be individuals, groups, communities or organisations. What all have in common is that they *act*. Their actions – or their intentional inactions – both inside and outside the boundaries of formal development initiatives shape the nature of change.[18]

There are three key points that anthropologists regularly and consistently make about development actors, as follows.

The first point is that development actors are *diverse*. People are all different, and so they engage with development in different ways. Specifically, development actors have different *social positions* and different kinds of *social roles*. Individuals may be male or female, wealthy or poor, educated or uneducated, old or young, from different ethnic backgrounds, different families and so forth. Groups and organisations may be large or small, formal or informal, influential or marginal, and exhibit various other social characteristics.

Their different social positions and social roles affect the nature of the resources available to development actors, and the amount of influence they have when it comes to negotiating change processes. Different amounts of resources and influence manifest as power differences. Some individuals, groups and organisations have more resources – money, land, political contacts, social networks, time, information, confidence – and others have less. Some people and organisations are influential – they are credible, respected and deferred to – while others have little or no influence. Because of power differences, some development actors will find it easy to be heard and get things done, while others will find it almost impossible. Some actors will lead change; others may be forced to follow. Because people are diverse, development processes will affect them – and be affected by them – in very different ways.

The second point is that development actors are *savvy*. Within the scope of the resources and influence available to them, people will generally do what they can to advance or protect what they care about. They are unlikely to do nothing at all, or to act against their own interests. Their scope for action may be limited by their circumstances, but they will still try to position themselves as best they can. They may act in ways that appear irrational to outsiders, but there is generally a reason for what they do. Anthropologists recognise that even when people do not act in the way that outsiders would expect, their actions nearly always make sense in the context of their own situations.

Anthropologists' up-close view of development actors shows that few people are truly passive – even the very poor have aspirations and create actions for change. Of course, sometimes the risks of acting are too high, or resources are simply unavailable. But temporary inaction is not the same as passivity. Anthropologists' observations of real people in real places demonstrate over and over that people are savvy: within the constraints they face, they find ways to create, support, resist or adapt to change – or to turn change processes to their own ends.

The third point that anthropologists regularly make about development actors is that they are *situated*. People do not live and work in a vacuum; they exist in a physical and social context. Development actors are located in real, on-the-ground places, and the resources available to them are largely defined by where they are situated: the nation they live in, the local environment they work in, and the social meanings that are attached to their gender, their family background, the organisation that employs them and so forth.

These contexts in turn affect whom development actors interact with, what resources they can mobilise (and at what scale) and, ultimately, what is – or is not – feasible for them to do. Some anthropologists refer to this as their 'room to manoeuvre'.[19] The context in which actors are situated affects their room to manoeuvre, be it in a large bureaucratic organisation or a remote tribal village. Further, people in context are persistently three-dimensional. They do not divide their lives neatly into silos with names like 'health', 'housing' or 'livelihoods'. Development actors are situated in contexts where different development concerns are frequently intertwined.

These three ideas about development actors are central to anthropology of development theory. Anthropologists recognise that development always involves people. People are diverse – with different identities, roles and resources. They are savvy – they act, with an eye to what they want and what they can achieve. At the same time, they are also situated in particular contexts which affect their ability to act and the strategies they are able to employ to create, manage or resist change.

Actors on the margins

Anthropologists have traditionally taken particular interest in development actors who are located outside centres of power – that is, so-called marginalised groups. These groups may be physically marginalised: located in remote or unpopular places such as jungle settlements or urban slums. Or they may be socially marginalised: members of minority ethnic groups or welfare recipients, for instance. In many cases they are both.

Marginalised development actors often end up on the receiving end of development initiatives designed by other actors: such as central governments, aid organisations or NGOs. The difficulty arises when those who design policies, programs and projects to benefit marginalised groups have little understanding of who these development actors actually are.

Marginalised groups are usually physically and socially distant from those who make the decisions. As a result, decision makers often know very little about them. Even when development initiatives are designed based on cutting-edge policy or scientific insight, they generally have very limited understanding of the actual situations of the people they are trying to help.

Too often decision makers see homogeneous 'poor' or 'disadvantaged' people, and fail to recognise that marginalised groups and their members are diverse: different genders, different ages, different social roles. Too often they assume that people living in difficult circumstances are helpless or ignorant, rather than savvy. They often do not understand that disadvantaged groups are situated in contexts that are very different from the ones in which decision makers operate, and which severely limit their room to manoeuvre.

Stereotypes and assumptions about marginalised groups can easily fill the gaps in what decision makers actually know about the people they are trying to help. From the receiving end, plans designed to make things better often look surprisingly uninformed. It is astonishing the things that development projects designed from afar can overlook: that a 'village community' has political and economic factions, for instance; or that 'poor women' have busy schedules.

Development initiatives designed as if all villagers had the same resources and interests, or as if poor women had plenty of spare time to take on new tasks, can become deeply problematic in practice. They reveal deep misunderstandings of what life on the margins is like. Failure to understand the realities of how people live and work can mean that initiatives designed to reduce poverty can actually create or exacerbate it.

On the margins, development actors are incredibly diverse. Yet policy makers often miss this simple point. Street children or members of a remote tribe are so different and so far away from decision-making centres, so it is easy to assume that they are all essentially the same. This is a form of stereotyping: imagining a typical 'street kid' or tribesman'; overlooking the fact that these groups are made up of individuals with different genders, ages, roles, ranks, resources and interests.

Decision makers are often caught by surprise when 'poor people' ignore or even oppose their development efforts. Yet development actors are savvy, and make their own judgements about what will benefit them. Decision makers tend to assume that marginalised groups are passive, sitting idly and waiting for a development program to appear, rather than – as is more often the case – actively managing competing priorities and demands: family, work, safety, food, relationships, social obligations. The process of defining certain groups as 'marginalised' focuses on their deficits, rather than recognising that they are often savvy operators in very challenging situations.

Yet these situations are, themselves, often invisible to outsiders. Decision makers seldom really understand the situations that marginalised groups face: the everyday realities of poverty, racism, insecurity or isolation. These situations are often well outside their life experiences. Nevertheless, these situations deeply affect how people who live in them see the world, and the room to manoeuvre they have to initiate or respond to change efforts.

Anthropologists present a very different view of development actors on the margins. Unlike most decision makers, anthropologists have worked up-close with marginalised groups: spending time with them, learning about their lives up-close and seeking to understand how they themselves see their world. Anthropologists recognise that development actors on the margins are diverse, savvy and situated. They are real people in real situations who act to protect what matters to them, and to make things better when they can. They seek change that meets their needs. They resist change that does not. In often very challenging situations they act within the options available to them.

Further, anthropologists show us that marginalisation is not simply a given: there is always a back story. People are generally on the margins because they have been placed there: removed from land, deprived of status, exploited as cheap labour over generations. Anthropologists remind us of the historical contexts that have made certain development actors marginalised to begin with, such as colonisation, dispossession or forced relocation. These historical situations often echo in a lack of resources and influence for actors from certain groups, right into the present. When trying to tackle today's development problems, anthropologists remind us that history matters.

Anthropologists as development actors

Anthropologists' interest in marginalised groups dates to their early tradition of work with communities in distant parts of the world. Anthropologists have crossed significant social divides – of race, class and nationality, for instance – to live with very different groups of people and study their ways of life up-close. As a result, anthropologists often have a very detailed view of groups of people that most other people know nothing about. When decision makers have sought to work more effectively with these groups, anthropologists have often been called upon to help.

As a result of their ability to navigate and communicate across social divides, anthropologists themselves have become important development actors. They have frequently been asked to mediate the connections between marginalised groups and external actors such as governments, aid organisations, NGOs and private companies. Anthropologists have done so at the request of both powerful organisations and marginalised groups themselves. Anthropologists' dual positioning as professionals (often from dominant social groups) who are known and trusted by marginalised communities has positioned them ideally to work with both mainstream organisations and marginalised communities.

Anthropologists have played active roles in development initiatives, both as practising development anthropologists and as engaged development scholars. Some anthropologists have worked for development organisations to help these organisations understand marginalised groups better and engage with them more effectively. Other anthropologists have worked for marginalised groups: for tribal bands, Indigenous land councils, peasants' unions and so forth, on a paid or voluntary basis, to help them negotiate with and gain support from powerful outsiders. Yet other anthropologists have sought to influence public debate and understanding about marginalised groups and development processes.

Anthropologists have acted to promote, mediate and/or resist development efforts, depending on their interests, beliefs and the situations in which they have found themselves. Like other development actors, anthropologists are diverse, savvy and situated in particular contexts that influence what they can or cannot do. Most frequently anthropologists have involved themselves in development through a desire to advocate for the interests of marginalised groups – groups with whom they have often worked personally – and to communicate the experiences and perspectives of these groups to developers and to the public at large. Often, their concern has been to ensure that development efforts benefit – or, at minimum, do not harm – groups on the margins (see Box 1.4).

Anthropologists have strong skills in moving across boundaries and understanding 'how things work' in very different social settings. They regularly find themselves at the meeting points between vastly different social contexts: remote communities and government departments, urban slums and international aid agencies, Indigenous communities and dominant post-colonial societies. Over the years many anthropologists have used their skills and knowledge

---

Box 1.4: Speaking for the 'victims of progress'

Anthropologists have often had a very ambivalent relationship with development, because they have seen how development initiatives have affected people – especially marginalised groups – on the ground. Some anthropologists have made a particular effort to write work that communicates the perspectives of marginalised groups vis-à-vis development to a broader audience. For instance, John H. Bodley's *Victims of Progress* (1990) describes the negative impacts of modernisation-focused development efforts for tribal peoples around the world. This book highlights the risks posed to culturally diverse groups and their local autonomy by externally imposed visions of development. Development initiatives often have losers as well as winners; even in the twenty-first century, development resource conflicts and economic pressures continue to place local livelihoods at risk. Anthropologists research the varying experiences of development and advocate for the interests of those who are placed at a disadvantage by current practices. Some anthropologists have come together to form advocacy organisations, like Cultural Survival (www.culturalsurvival.org) which works to promote respect for Indigenous peoples' worldviews, lifestyles and rights.

---

in boundary-spanning to help very different development actors to build relationships with each other. As brokers, they have helped less powerful groups to communicate with and understand powerful outsiders. And they have helped powerful outsiders to communicate with and understand distant and unfamiliar groups. Often they have worked to translate the diversity, savviness and situations of local people in a language that very differently situated professionals can understand.

Anthropologists' experiences working across social divides, with both marginalised and powerful groups, have helped them become skilled brokers and advocates. Nevertheless, these kinds of roles are less relevant than they once were. As members of marginalised groups have become more politically aware and confident to speak for themselves, there is less of a need for professional advocates and brokers to speak for them. Still, anthropologists regularly take their boundary-crossing skills into other domains of development work: engaging with multiple disciplines on project teams, or working at the meeting points of diverse communities and policy bodies.

Most anthropologists still align themselves strongly with the interests of the least powerful development actors – those on the receiving end of development initiatives. This attention to local people and their views does not always sit comfortably with development professionals, however. Anthropologists' strong attention to 'social' and 'community' issues has often meant that their work is seen as less valuable than more technically oriented work. It is frequently positioned as marginal to the perceived core business of a development program or policy.

Brokering relationships across boundaries and advocating for less powerful groups are not easy tasks. Yet these are tasks that anthropologists, as development actors, often undertake. This work is not always influential, and powerful decision makers may wonder why they should care. Yet when it comes to development, anthropologists have been in a privileged position, able to understand 'both sides of the story': the perspectives of marginalised groups and the perspectives of developers themselves.

Developers as social actors

Anthropologists are particularly known for their ability to provide insights on the perspectives and experiences of marginalised groups. These are the traditional *developees*, the so-called 'target groups' of development action.[20] Increasingly, however, anthropologists who study development have widened their scope beyond developees to include a range of other relevant development actors. In particular, many have become interested in those who aim to drive the development agenda: the developers.

In recent years, anthropologists who study development have turned their attention to understanding developers themselves. Some refer to this as 'studying up': that is, studying powerful organisations and how they work. *Developers* may be large organisations with complex ways of working – such as national government ministries or the World Bank. Equally, however, they may be smaller organisations such as grassroots NGOs or local progress associations. They may also be the individuals who work in these organisations, as paid professionals or as volunteers. An entire sub-genre of anthropology of development literature now looks up-close at the lives and work of international aid workers.[21]

In studying the workings of development organisations and the people in them, anthropologists have mobilised their ethnographic skills: specifically, their ability to view social situations as they unfold on the ground and from within the perspectives of different social groups. However, instead of heading off to far-flung villages to work with the recipients of development assistance, they have immersed themselves in the social contexts of boardrooms, head offices, and national and international centres of decision making. Or, alternatively, they have approached the far-flung villages to study what developers do there.

Anthropologists have been able to observe up-close the social processes through which developers 'do' development: in high-level fora, national parliaments, organisational offices, and in particular development programs and projects rolling out in particular settings all around the world. They have observed development actors in a wide range of organisations, from major multilateral and bilateral organisations to local government councils and NGOs. Even when development looks like a highly technical process, anthropologists have shown how the process is negotiated among development actors – who are, as always, diverse, savvy and situated.

Anthropological studies of development practice reveal the presence of diverse developers – from directors of large organisations to front-line workers. Developers have different social positionings and roles. Some control considerable resources and influence, others very little. Individuals and organisations have different values and motivations for their involvement in development. Like other development actors, developers are savvy in pursuing their particular agendas – including their own visions of what good development looks like. Importantly, developers are also situated in particular social contexts that affect what they are able to achieve: the organisational contexts in which they work, with their structures, processes and hierarchies, as well as external political, institutional and community contexts.

No two developers are the same. Developers are diverse, and situated differently. A government ministry has more influence than a small NGO – though the NGO may be more agile on the ground. The head of an organisation has more influence than a project officer – yet it is the latter who often has the direct relationship with project 'target groups', and is therefore more likely to know what will actually work. Males generally have more influence than females; professional staff generally have more influence than volunteers; federal ministers have more status than local councillors; and expatriate staff tend to earn more and stay in nicer hotels than local staff do. These social positions matter. Developers have different social roles and positions, which gives them different amounts of room to manoeuvre when it comes to creating change.

Further, developers are nearly always situated within organisational frameworks that determine what can and can't be done. Development organisations tend to run in certain ways: they have program areas, strategies and plans, budget cycles and funding deadlines, reporting requirements and review processes. Government-based organisations are framed by their political environment and the concerns of ministers and big-picture policy directions; NGOs are attuned to funding sources, core mission and stakeholders. Most change-creating initiatives are designed and implemented in recognisable forms – such as projects and programs, rolled out in structured cycles. These determine how the organisation – and the people in that organisation – work. As anthropologist Judith Tendler discovered when she looked up-close at USAID, develop ment organisations work in particular ways, and these ways of working shape constraints and incentives for those who work in them.[22]

Within the possibilities open to them, developers have diverse agendas and aspirations which they aim to progress. Organisations have their guiding policy directions and mission statements – but they also have stakeholder pressures and political expediencies. Individual development workers have values and personal missions – but they also have position descriptions, work plans and career goals that shape their choices. Different developers want different things – and, indeed, may want different things at different times.

Further, developers must always negotiate with others – other developers, stakeholders and would-be developees – to make development happen. As David Mosse has shown in his work on *Cultivating Development*, looking up-close at the work of development organisations reveals that myriad negotiations are required to create a project. Projects are never simply 'rolled out'; rather, the actors in charge of implementation must continually negotiate with other actors to create the project and its results.[23] As Emma Crewe observed recently in a case study of an international NGO, planning and policy did not equate to practice; when development plans and policy ideas hit the ground in real social contexts, they needed to be represented, communicated and negotiated.[24] Development initiatives are always negotiated and, ultimately, co-constructed by different actors.

Understanding development as *something people do* is the first step toward an anthropological understanding of development. Understanding how development projects, programs and policies work – and how to make them work better – means paying attention to development actors. Development actors are diverse: many people and organisations play a part in instigating, promoting, managing or resisting change processes. These include developers, developees, other stakeholders and those whose roles cross over, such as people or organisations that instigate projects for their own communities.

Anthropologists recognise that development actors have different social positions and roles, different situations and, often, very different agendas. They show how development happens in their interactions. And they remind us that the results look different depending on who you are.

Development knowledges: What people know

For anthropologists, development is something that people do. Observing how development plays out in real social contexts reveals that a wide range of people and organisations are involved in negotiating change. Anthropologists pay attention to the on-the-ground realities of these diverse actors. They pay attention to their different roles, resources, situations and development agendas. Then, they look deeper.

Anthropologists seek to understand development as it happens on the ground in real places, but they also aim to understand it *from within* – that is, from within the perspectives or viewpoints of the people involved. Understanding development 'from within' means understanding how the worldviews of development actors influence what they do. It means paying attention to how development actors see the world, what they know about it and how they organise themselves in response. In short, understanding development from within requires attention to *culture*.

Culture is a word that has been used a lot in anthropology. It can be defined as the ways of knowing, thinking and acting that are broadly shared by members of a social group. This does not mean that every member of the group knows the same things. It does not mean that everyone in the group thinks or acts in the same ways, or believes the same things as everyone else. Culture is a framework rather than a box. It is a set of *shared ideas and practices*, with the potential to vary, rather than a fixed container that everyone fits into.

Thinking of culture as a 'framework' or way-of-framing is useful. *Frames* influence what we see: such as the framing of a picture, or the framing of a window. Shared ideas, ways of thinking and ways of acting 'frame' how people understand a situation – what they see as important or not important – and how they respond to it. In development work, a large global organisation like the World Bank will generally see development problems very differently than a small Indigenous NGO. An organisation working in the health sector will see development issues differently than one working in microfinance. Their frames are different.

Cultures – the shared ideas and practices within a social group – influence how members of a group frame problems and solutions. Culture frames what they see as desirable or undesirable, and what they see as possible or impossible. Yet these frames are neither permanent nor static. In a different social situation, with a different group of people, the same person may frame things differently.

What is seen as appropriate behaviour or an acceptable reason in one context may be seen as inappropriate or unacceptable in another; what is surprising in one setting may be unexceptional in another.

For a long time anthropologists thought of culture as something that was static and easy to define. There was a tendency to see culture as something a group of people *had*, or something to which people *belonged*. Cultural character-istics were assumed to be largely fixed, remaining the same over generations, so that it was possible to speak authoritatively of a particular culture and how it worked. Change was happening, but it was often simply overlooked. The kilt of the Scots or the bowler hats worn by Aymara women in Bolivia are of relatively recent historical vintage, yet they are often portrayed as core markers of the 'traditional culture' of these groups.

Contemporary anthropologists recognise that culture is neither fixed nor static. People are always meeting other people, learning things and trying new things out – and have done so for generations. Cultural frames provide people with a 'repertoire' of potentially acceptable understandings and responses, which they may choose and combine as required. Further, 'traditional' cultural ideas are re-applied in new ways, extended and reinvented all the time. People regularly challenge established cultural ideas (think of abolitionists challenging slavery, or revolutionaries challenging monarchies). And people regularly try out new practices that press at the comfort zones of what is considered 'culturally acceptable'. As a result, normal behaviour or accepted practice changes over time.

Thus, while it is true that core cultural ideas often persist over generations, many of the trappings of culture – dress, language, social categories, daily practices – are continually in flux, negotiated, stretched, challenged, affirmed and periodically reinvented. Because culture is so fluid and negotiable, many anthropologists have become wary of the term *culture* altogether. They do not want to give the impression that they are talking about a fixed, static thing. They recognise that people are not locked into a single cultural identity, or defined by an unchanging set of cultural characteristics.

Culture is fluid and negotiable – but it is still important. The shared ideas, beliefs and practices of a group of people are often core to community identity. They are the frames that people use to make sense of the world, and to organise and explain to themselves and others their actions in it. Even when people are not aware of them, these cultural frames guide how members of a group interact with each other and with those outside their group: why they see some people as 'important' and others as unimportant, some practices as 'normal' and others as problematic, and some outcomes as 'desirable' and others as undesirable. People are not locked into their culture, but they are always guided by cultural ideas, both those of recent vintage and those that have been handed down through generations.

When it comes to development, culture influences development work in two key ways. First, culture frames what people know about their world and how they know it. Knowledge in development takes many different forms, from the evidence base of a large development program to the indigenous knowledge of traditional healers. Each kind of knowledge is based in deeper cultural ideas about what is valued and what matters – and therefore what is important for people to know. These cultural repertoires of knowledge – general and specialist, shared or contested – enable people to understand and visualise what is possible, what is desirable and what change is likely to mean for them. Equally, they provide a resource that people can draw on to shape development debates.

Next, culture frames what people do and how they do it. Shared ideas and beliefs lead to shared practices, such as the writing of project proposals or the conduct of cleansing rituals. For any given group of people, certain practices will be comfortable and well-accepted; other practices will be challenging, foreign or problematic. In development work, different groups of people may work in surprisingly different ways: some via formal meetings, others via informal gatherings; some via extended consensus-building, others via executive decisions. Ways of working matter in development work because groups of people may be offended, alienated or even directly excluded by others' ways of working, or may feel pressured to work in ways that are uncomfortable or disempowering for them.

This section introduces ideas from anthropology about *development knowledges* – in the plural. It explores development actors' different ways of knowing and thinking, and why these matter for effective development. From there, the following section introduces ideas from anthropology about *development institutions* – with a small i. That section explores development actors' different ways of working and why these matter in development work. Together, these two sections deepen the idea that development is a social process, by showing that it is not only social but deeply cultural: a process in which different ways of knowing, thinking and acting come together.

The idea of development

Knowledge is built on shared concepts and ideas. What people know, and how they know it, is ultimately cultural. When it comes to development, cultural ideas affect how people understand positive change and what they believe is required to achieve it. For some, *positive change* may be ultimately spiritual; for some, eminently practical; and for others, it may be a contradiction in terms – why would we wish to change, when we like things as they are?

Different groups of people have different understandings of concepts like *positive*, of ideas like *change* – and, especially, of historically powerful ideas like *development*. While this book defines development as the process of social and economic change – and development practice as the intentional pursuit of positive change – there are many definitions of what development is and many views on what 'positive' might entail. All around the world, people have different understandings of development which are grounded in their own ideas and beliefs about the change that they want (or do not want) for themselves and their communities, as well as their ideas about the nature of the changes that others may or may not want for them.

As an example, the Khumi people in the hills of Bangladesh have their own local idea of development which they call *nebu-heina* or better-life. Anthropologist Nasir Uddin has documented that for the Khumi, development is better-life that selectively includes some aspects of non-Khumi education, lifestyles and livelihoods, but still remains firmly grounded in the values and lifestyle of Khumi communities. This view of *development* is framed within the Khumi social and cultural context; as a result, it is quite different than the view of *development* found among the non-Khumi educated classes of Bangladesh.[25]

Similarly, in Malaysia anthropologist Anne Kathrine Larsen has documented two very different concepts of development: one that is based in externally imposed ideas about modernisation, and one which is a local, holistic concept of a good life. She observed that in Malaysian villages there was a strong distinction between the ideas of 'not yet developed' – synonymous with backwardness – and 'developed' or modern. Alongside this sharp dichotomy, however, there was another, distinctive local concept of development: *hidup sempurna*, which means the 'perfect or complete life'.[26]

In Ecuador and Bolivia, indigenous Quechua and Aymara concepts of living well or the good life (translated into Spanish as *buen vivir* or *vivir bien*) are being employed as an explicit alternative to mainstream Western ideas about development. *Buen Vivir* (in Ecuador) and *Vivir Bien* (in Bolivia) have been used by governments in these countries to explicitly articulate indigenous Andean ideas about well-being and position these ideas at the centre of national development policy. *Buen Vivir* and *Vivir Bien* seek positive change with a holistic and community focus, in contrast to more economic and individualistic definitions of development. The attempt to institutionalise these ideas in national policy environments is an interesting story in its own right (see Box 1.5).

In Tanzania, development is *maendeleo*, 'going forward'; this idea is 'deeply embedded in notions and images of change, modernity and prosperity' including education, knowledge of technology, religion, cleanliness, housing, dress, food and so forth.[27] In many parts of the world, these kinds of ideas about modern lifestyles, and an imagined linear trajectory from 'less' to 'more' developed, continue to infuse how many people think about development. These ideas regularly portray 'modern' housing, infrastructure, dress and jobs – often imported from elsewhere – as 'developed'. Local constructions and lifestyles, however practical or meaningful, tend to be framed as not-modern, and thus symptomatic of less development, even backwardness.

Anthropologist Arturo Escobar has called attention to the problems that result when people buy in to ideas about development that others impose on them. Specifically, he documented how Western ideas about *development* have become dominant and have created serious problems for the rest of the world. When *development* is equated with modernisation, capitalism and the trappings of Western lifestyles, a false dichotomy is created between places that display Western lifestyles and values and those that do not. The former are defined as *developed*, while others are devalued: portrayed as lacking, *underdeveloped* or *developing*. Escobar has shown how the language used in international development work perpetuates certain dominant ideas, beliefs and worldviews that essentially *create* underdevelopment: by defining development in certain ways – and then exhorting others to catch up.[28]

It matters deeply how development is defined and who gets to define it. Development organisations have their own cultures, and thus their own ideas about what development is. While these ideas clearly vary from organisation to organisation, there are some common patterns: development organisations tend to frame *development* as something that they can do something about. In a famous case study of development work in Lesotho, the anthropologist James Ferguson demonstrated how development organisations viewed *development* as a problem that required the kinds of technical solutions that the development organisations could provide. By focusing on technical problems while ignoring political drivers of change, the development industry in Lesotho systematically overlooked many of the real, political reasons why change was difficult to achieve.[29]

Some development organisations define development quite narrowly, others broadly – depending on their guiding ideas and how they frame positive change. In local government councils in Australia the language of *development* refers first and foremost to planning applications; thus, development is still seen largely as a question of planning for the correct infrastructure. Simone Abram has observed a similar preoccupation with buildings and infrastructure in development work in the UK. She observes that despite efforts to encourage public participation in decision making about future development, the actual scope of participation is severely limited because 'development per se is always assumed' – and development means infrastructure. The scope for participation 'only touches on its location, and some aspects of its design'; the idea that participants may not want development is, for the planners, 'unthinkable'.[30]

The moment of defining *development* can indeed be a defining moment – particularly if powerful development actors have the resources and influence to ensure that their definition prevails. An anthropological case study from Malta provides a vivid example. The authors document how government actors and local NGOs defined *development* quite differently – and what happened when these ideas about development came into conflict. Government actors in Malta viewed development in economic and political terms, as something to be achieved via employment and high-profile, upscale tourism projects. Local NGOs, by contrast, emphasised social and cultural values as central to successful development. When a large hotel development was proposed, NGOs' concerns about the loss of social and cultural amenities were ignored. The ideas about development that were shared by planners, bureaucrats and developers supported the hotel, so they mobilised their resources and influence to rush the project through. The NGOs' views about development were dismissed as 'emotional and subjective, thus unscientific'.[31]

When development is defined in a particular way, other ideas about positive change easily become invisible. Some actions start to look logical and inevitable, while other possibilities fly under the radar. Certain types of knowledge – like tourism figures or annual turnover – become valued and highly visible, because they are grounded in dominant ideas and support dominant views. Other types of knowledge are invisible or dismissed as irrelevant to the development task at hand. Anthropologists' up-close view reveals that different people have different ideas about development – and some of these are more visible than others. The ideas and knowledges of some development actors are regularly privileged, while others are overlooked, misunderstood or even actively suppressed.

Indigenous knowledges

In the same way that anthropologists have taken a particular interest in marginalised social groups, they have also taken a particular interest in marginalised knowledges. Because anthropologists have studied different groups of people around the world up-close, they have spent time learning about what people in these groups know. They have observed that this knowledge is important in how people live, and indeed directly relevant to change efforts. When anthropologists have become involved in development work, they have called attention to the knowledge that these marginalised groups already have about the development issues that outsiders seek to address.

One of the most common mistakes in mainstream development work is to overlook the knowledge of the people who are on the receiving end of development initiatives. Too often, it is assumed that these groups, by virtue of their marginal status, do not have *knowledge* – because they are not educated in mainstream education systems, cannot communicate in the dominant language, or because what they know does not look like what outsiders expect.

Back in the 1980s, Robert Chambers exhorted development professionals to pay attention to 'rural people's knowledge', particularly, to take notice of what rural people themselves know about rural development.[32] There have always been some development professionals who have been attuned to local people's knowledge, but they have been in the minority. In recent decades, however, there has been growing awareness in development circles that marginalised groups have knowledge that is relevant to development work. Nevertheless, there has been considerable confusion about what this means in practice. For the most part, the focus in development practice has been on asking people's opinions rather than recognising their knowledge. Where knowledge has been recognised, it has been limited to a handful of specific domains of technical knowledge, such as farmers' knowledge or local environmental knowledge.[33]

Anthropologists have had a much deeper and longstanding engagement with the knowledge of marginalised groups. They have attempted to bring this knowledge into dialogue with development practice in a range of ways. Among anthropologists who have engaged in development work, *indigenous knowledge* has been a key topic. Indigenous knowledge can be defined as the on-the-ground, culturally situated knowledge of particular social groups, especially (but not exclusively) those that have a longstanding presence in a particular area. Indigenous peoples, as the original inhabitants of land that was later colonised by others, are the most obvious possessors and practisers of indigenous knowledge. Yet indigenous knowledge is a broader term, and may refer to the internally shared knowledge of other social groups as well.

Indigenous knowledge has been called by a number of terms, including traditional knowledge, local knowledge and community-based knowledge. All broadly emphasise that this kind of knowledge is *situated* in particular physical and social contexts. What characterises indigenous knowledge is that it has been developed in the particular context where a particular social group lives, often over a considerable period of time – indeed, for some Indigenous communities, over millennia. Indigenous knowledge can be thought of as *situated knowledge* that has been developed in particular physical and social settings – as opposed to abstract knowledge, which can be generalised across settings.

A number of anthropologists have focused on studying indigenous knowledge and showing how it is deeply applicable to development work. Early work by anthropologists including David Brokensha and Michael Warren launched the field of indigenous knowledge studies in development.[34] This work focused heavily on documenting indigenous technical knowledge (ITK). ITK is situated knowledge that is directly and practically applicable; for instance techniques for farming, housing, caring for animals, sourcing and preparing food, curing disease, making handicrafts and so forth. ITK reflects people's knowledge of their own environments and the ways that they have engaged with them and adapted to them over time. Early studies of ITK paid considerable attention to detailed typologies of indigenous concepts, categories and environmental variables, and began to show the relevance of this knowledge to practical development work.

The next wave of work on indigenous knowledge, led by anthropologists Paul Sillitoe, Johan Pottier and Alan Bicker, explored indigenous knowledge with much more explicit attention to political context and values.[35] This work moved beyond descriptive inventories of indigenous knowledge and its relevance, to consider specific experiences of how indigenous knowledge is used in development work. Anthropologists explored how knowledge for development is negotiated across cultural divides, and how aspects of indigenous knowledge and values may be systematically misrepresented, overlooked and devalued – even in well-meaning and ostensibly participatory development processes. In addition, in recent years, there has also been greater attention given to intellectual property and the problems of appropriation of indigenous knowledge by outsiders.

ITK has become known in development circles to the extent that it has acquired its own acronym. Yet, other forms of indigenous knowledge are less well known. Technical knowledge has historically received the bulk of attention because it is comparatively easy to observe, and obviously applicable in development work. Yet in addition to technical knowledge, there are other kinds of indigenous knowledge that have been extensively documented by anthropologists, and which are also directly relevant to development work. For ease of reference, these can be divided into the broad categories of *experiential knowledge* and *cultural knowledge*.[36]

Experiential knowledge is the knowledge that people gain from experience. Attention to indigenous experiential knowledge opens the window on history, and begins to reveal, often in very concrete ways, how historical experiences influence current actions. For instance, in Malaysia, Larsen observed that, 'Villagers took up a waiting attitude to actual projects, especially income-generating ones, as they have witnessed many failures along that line.'[37] Experiential knowledge can also be developed through experimentation. Anthropologists such as Paul Richards were among the first to point out that farmers regularly experiment and try out new things – they 'perform' what they know about farming in their local environments, and in performing it, grow knowledge about what works and what doesn't.[38] These ideas about on-the-ground experiential knowledge influenced later development practices such as Farmers' Participatory Research.

Cultural knowledge, in turn, is the knowledge that people gain from belonging to a particular cultural group. Cultural knowledge includes the group's shared ideas about what things mean and how things work in a given social context. This kind of knowledge is expressed in a group of people's language, beliefs, values, rituals and the everyday details of social organisation, down to the practical questions of who lives where, who marries whom, and who gets to make the decisions. These decisions are not culturally determined, but they are informed by cultural knowledge about what other members of the group will accept. Further, cultural knowledge often has its own categories and frameworks – including its own *epistemology*, or ideas about what knowledge is. As Paul Sillitoe has written with reference to Indigenous scholarship: 'it challenges not only the propriety of the categories we use, but also. . . Western ideas of what knowledge comprises.'[39]

Cultural knowledge can be particularly difficult for outsiders to grasp. It may sit outside their own epistemologies or understandings of knowledge (for instance, people with scientific training may believe that all true 'knowledge' is abstract and generalisable). Yet cultural knowledge is important in practice, because those who are unaware of unwritten cultural rules can find social settings difficult or impossible to navigate. They can find themselves breaching social protocols and, as a result, not being taken seriously: speaking with 'leaders' who do not really lead, for instance, or suggesting courses of action that are offensive or counterproductive. Cultural knowledge is thus particularly important in development work, where different cultural groups often find themselves working together. Nevertheless, it is the least understood aspect of indigenous knowledge.

Indigenous technical, experiential and cultural knowledges are all relevant to development work. These three forms of indigenous knowledge explain a great deal about how things work or don't work in a given development context – well beyond the technical strengths or limitations of development initiatives. Despite this, development professionals have often failed to notice indigenous knowledge or take it seriously. Physical and social distance between decision makers and marginalised groups can mean that the former have little awareness of the knowledge of the latter. Indeed, sharp differences in education and social status often lead decision makers to assume that they are the only ones who have valid knowledge.

Decision makers regularly fail to take into account what local people know about their local physical, social and cultural contexts. The tendency to stereotype on-the-ground development actors as ignorant, or make assumptions about what they do or don't know, has often led to a situation which anthropologist Mark Hobart terms 'development ignorance'.[40] *Development ignorance* is the systematic suppression of indigenous, on-the-ground sources of knowledge by development professionals in their work. Development ignorance is usually unintentional, but it has real consequences in terms of the knowledge that is available for informed decision making. Even when decision makers recognise that indigenous knowledge exists, they tend to miss its nuances: acknowledging ITK, for instance, but not the cultural or experiential knowledge that frames technical decision making – particularly when this knowledge conflicts with decision makers' own worldviews.

Like development actors themselves, indigenous knowledge is *diverse* and *situated* in social contexts. Decision makers often assume that any member of a group has the same knowledge as everyone else. They thus ask one 'representative' of the group to speak for everyone. This is a mistake; one person seldom manages more than a portion of all the local technical, cultural and experiential knowledges present. In any social group, different people have different knowledge: characteristics such as age, gender, occupation or social class will influence who knows what. A young person will not know what elders know. Men and women will often manage quite different knowledge sets. The 'old guard' in an organisation will have different knowledge than newcomers.

Indigenous knowledge is always present in social groups that live and work together in particular contexts. In some cases, it is deep, rich knowledge with hundreds or even thousands of years of heritage. In other cases it is of more recent vintage, but still deeply situated in specific physical and social contexts. Outsiders commonly overlook or misunderstand indigenous knowledge. When these outsiders hold the power to make development decisions, their 'development ignorance' becomes dangerous. A failure to recognise what local people know can mean that developers undermine functioning ways of working and leave people worse off than before.

Development logics

Anthropology is the one Western science that has intentionally attempted to look beyond Western ways of seeing the world. It has sought to understand and appreciate the knowledge of other cultural groups. By seeking to understand the worldviews of different groups of people around the world, anthropologists have continually questioned our own assumptions about how the world works.

When it comes to development, anthropologists show us that our own ideas about what development is, and what knowledge is required to achieve it, are only one part of a bigger picture. We each 'frame' the world – and our ideas about development – in particular ways. Other people frame things differently – and their ideas may surprise or challenge us. By calling attention to what other social groups – especially those 'on the margins' – know, anthropologists reveal different views of development, and suggest previously unimagined development possibilities.

We have seen that around the world people have different ideas about what positive change is and how to achieve it. Anthropologists refer to these as *development logics*.[41] A development logic is a set of ideas about the kind of change (or non-change) that is desired and what is required to achieve it. These ideas are framed by culture – the ways of knowing, thinking and acting of each group.

The development logic of a large bilateral aid organisation – the way that members of that organisation define positive change and go about creating it – is usually quite different from the development logic of a small NGO. They may both use 'development words' like *poverty* or *participation*, but use them to mean different things. Both their approaches to positive change, in turn, are likely to be markedly different from the development logic of a rural household. Different groups, and indeed different individuals, have different development logics. They know and believe different things about the world; and this frames how they see and do development.

The anthropologist Jean-Pierre Olivier de Sardan distinguishes between *notional logics* and *strategic logics* in development. Notional logics are ideas about what things are: *What is positive change? What is development?* Notional logics are about how people *see* development. Strategic logics, by contrast, are action-oriented: *What should be done? How should we do it?* They are about how people *do* development. Olivier de Sardan makes the point that when different actors come together in development initiatives, they often have different logics – and this affects how they work together.[42]

The development logics of an NGO and those of a rural craft cooperative may be quite different even when they are working together on the same project. They may even use the same words – *success, empowerment* – but they have different notional logics about what these words mean. Further, they may have different strategic logics about what it is necessary to do to achieve their goals. The two organisations may encounter tensions when they try to decide the best course of action. They may find themselves pulling in different directions or being surprised by what the other is doing. When different groups come together to do development, they may not realise that they are operating according to different logics.

Because different development actors have different logics, their actions may not necessarily look 'logical' to one another. For instance, a community group may accept a project that they don't really want. They may do this because they know that having the project will strengthen their relationship with a powerful organisation – and this could be useful down the track. Any development professional who has spent time in the field will generally be able to provide examples of times when people or groups have done things that did not seem very logical. On closer look, however, these actions often have a logical explanation – one that made sense to the people involved.

When development actors act, their actions are guided by their own logics. Even when people are doing the same things, they may be doing them for different reasons. For instance, one individual may join a neighbourhood association because they believe that working together will allow the community to create more opportunities for young people. Another joins the same association because it will provide a platform from which to launch her political career – dedicated to helping young people. These two individuals have different strategic logics: the first sees that change will happen through community-based action, while the second believes that political power is required for change.

People and organisations thus 'frame' the development problem and solution differently, according to their own notional and strategic logics. These logics are cultural: based upon what people know and believe about the world in general and about the specific contexts in which they live and work. People from different cultural backgrounds, living in different contexts, tend to have quite different ideas about what 'development', the 'good life' or other positive change looks like. Equally, they have different ideas about what course of action is logical and possible in order to achieve the change they want. Nevertheless, decision makers tend to overlook or ignore development logics that do not match their own.

Anthropologists' up-close view of social settings reveals a multitude of knowledges and logics about positive change. Development actors in particular places know a great many things about their social and physical contexts that are not immediately apparent to outsiders. Looking at social groups 'from within' begins to reveal these knowledges. Culture frames what people know and how they know it. Because local frames and outsiders' frames are different, locals approach development differently than outsiders might expect.

For development professionals, recognising the existence of multiple development knowledges and logics can help to make sense of situations that look illogical. It can help to avoid costly errors borne of 'development ignorance', and tap into rich local knowledge sources to identify what works (and doesn't work) in particular settings. In the end, attention to the knowledges and logics of 'developees' provides important guidance about what is actually needed to achieve effective development solutions.

Development institutions: What people do

In colloquial language, *development institutions* are big development organisations, like the World Bank or the United Nations. Academics, however, have a more precise definition of institutions. Institutions are not the same as organisations. Rather, *institutions* are ways of working. They are the structures, rules and norms that shape how things are done. *Organisations* are just one kind of structure that people use to get things done. Thus, institutions such as international development cooperation can be formalised into organisations such as the United Nations or the World Bank.

Organisations are concrete things: organised groups of people. Institutions, by contrast, are hard to see: they are ideas and practices. Institutions are thus deeply cultural; they are part of a social group's shared ways of knowing, thinking and acting. Specifically, institutions guide how people choose to act, and they underpin the ways that people organise themselves.

Paying attention to how institutions work can tell us a great deal about social and economic change processes. Institutional economists like Douglass North have argued that institutions – structures, rules and norms – matter for economic outcomes.[43] Anthropologists use the language of 'institutions' less often; nevertheless, they have long experience studying the ways of working of different groups of people. They have shown how people in different parts of the world use different structures, rules and norms to organise their social and economic activities.

This up-close look at social contexts reveals that even 'economic' activities are deeply social. *Economic anthropologists* are a group of anthropologists who have been particularly interested in how economies in different parts of the world work. Economic anthropologists study how people in different places organise their economic activities: work, business, trade and so forth. These activities look different in different places. People have different ideas about what is valuable, what constitutes work (and who does what kinds of work), ownership (what can and can't be owned, and by whom), and the appropriate and fair processes for trade, exchange, gifting and so forth. In different social settings, there are different structures, rules and norms that guide economic activities. Economic institutions work differently in different contexts.

'The market', for instance, is an institution; it is a way of doing exchange. While there are 'global' market institutions that are shared across national boundaries, there is no single *market*. Anthropologists have shown that markets work quite differently in different places around the world. There are different rules and norms about what can and cannot be exchanged, where and under what terms. People trade in different places, in different organisational settings and make use of different structures to facilitate or limit economic activity. For instance, in some markets, long-term social relations are key to success; in others, short-term competitive behaviour is the way to get ahead. From sprawling open-air markets, to multi-storey shopping malls, to websites with e-commerce facilities, there are many ways of doing exchange. And all of these ways of working are deeply social.

When it comes to understanding – and enabling – change, it is important to understand how things currently work. Attention to institutions-with-a-small-i calls attention to people's current ways of working. Anthropologists reveal the variety of economic and social institutions around the world – not just those we know well or those that we are comfortable with, but also those we fail to notice, or which may operate in surprising ways. Anthropologists show us that institutions shape how people work and act. Thus, they shape how change happens – as well as why things stay the same.

Culture and institutions

Institutions – people's ways of working – may not look like we expect. Even familiar institutions, like markets, school systems or health services, may look very different in different places. Anthropologists have studied the ways of working of exotic and unfamiliar social groups in distant places and described, in ethnographic detail, the structures, rules and norms that people use to organise themselves and their activities. More recently, they have used the same techniques of observation and explanation to understand how more familiar communities and organisations work. While they may not look very exotic on the surface, a corporate office, factory floor or community organisation often has surprisingly complex structures, rules and norms governing how things are done.

In *The Lobster Gangs of Maine*, for instance, anthropologist James Acheson analysed the economic activities of lobster fishermen in the North-eastern United States. His work shows that a complex web of social relationships underpinned lobster fishing as a successful economic activity. In this context, 'survival in the industry depends as much on the ability to manipulate social relationships as on technical skills.'[44] The ways of working of lobster fishermen were underpinned by particular structures, rules and norms that outsiders would struggle to understand, but which were central to success in this economic activity.

Rather than using the language of *institutions*, anthropologists have most often referred to these structures, rules and norms simply as expressions of *culture*. Structures, rules and norms ultimately come back to shared ways of knowing, thinking and acting: how a group of people sees the world, and how they organise themselves in response.

Anthropologists have documented the different institutions that people around the world use to do all of the things that development organisations are typically interested in: to produce and exchange goods; to lend and borrow money; to educate children; to prevent and heal disease; to protect vulnerable members of the community; to manage environmental risk; to gain the support of powerful allies; and so forth.

Working around the world, from remote Amazonian tribes to large corporations, anthropologists have documented an abundance of ways that people do all of these things. People organise themselves into structures (like households, schools or workplaces) that are characterised by rules and norms about what should be done, by whom, where, when and why. These rules and norms often position different people differently: women in a different situation than men, for example, or natives in a different situation than foreigners. Structures, rules and norms frame what can be (acceptably) done and who can (acceptably) do it. They therefore directly influence people's 'room to manoeuvre'.

Anthropologists have described how institutions vary from place to place, from one social context to the next. There are many different institutional arrangements around the world for organising families, work, leisure, education, exchange, resource management, social safety nets and political action. Even apparently similar institutions – such as 'primary schooling' or 'community banking' – look very different in different places. They have quite different structures, rules and norms, and these affect people's room to manoeuvre in different ways.

Take a common institution like the household. In development work, the *household* is an important unit of analysis. Often projects are designed to involve and benefit households, and impacts are measured at the level of households. 'Household' seems to be a common structure and organisational principle all around the world. Yet around the world, households are quite different.

In some places, households are organised as nuclear families (one or two adults and their children); in other places they are organised as extended families, with multiple adults and multiple generations. Equally, households may include people who are not related at all: friends, housemates, godchildren. When people marry, the norm may be for the new couple to live near the husband's family (*patrilocal*), the wife's family (*matrilocal*), or to live away from the extended family altogether.

The roles of household members also vary from place to place. Anthropologists remind us to look at the household up-close and remember that household members have different roles and situations. These vary according to gender – whether one is a man or a woman – and according to age – whether one is younger or older. Social rules and norms influence the types of work that men and women, older generations and younger generations within the household do, their access to resources and their influence over decision making. The nature of these institutional arrangements varies from place to place. Equally, changes to household activities – such as those introduced by a development project – may improve or exacerbate the situations of household members in certain roles, such as women's access to resources or decision-making power.

Institutions are shared ways of acting, and they vary in often subtle ways from place to place. Because of this, it is dangerous to assume that we know what a household is – or what it should look like. It is dangerous to assume that we necessarily understand the nuanced internal ways of working that determine who within a household or a community can access resources and influence, and who cannot. Robert Netting notes that anthropologists are skilled at describing the 'actual functioning of real systems' on the ground, showing both 'how existing local institutions work and their potential for change'.[45] This is an important point: to see the potential for change, it is first necessary to understand how institutions already work on the ground.

Nevertheless, institutions on the ground can be difficult to see and understand. For outsiders that 'frame' the world according to a different set of ideas and practices than the locals do, it can be difficult to understand how other people's institutions work, or the functions they perform. Many institutions around the world perform practical and/or symbolic functions: building consensus, minimising risk, assisting redistribution or maintaining social order, for instance. Nevertheless, these logics are not necessarily immediately obvious to an outsider.

It is therefore easy for outsiders to miss the institutional clues, and run afoul of 'how things are done' in a particular social context. The institutional rules and norms that guide how things work are, after all, generally unwritten. The structures are not immediately obvious. Knowledge of how things work in a particular social context is cultural knowledge. Those who lack it may find it very hard to get things done.

Institutions and development

Institutions are central to development practice. The starting point for any action for change is the present: How do things work now? If the issue is education, what institutions are used to provide education? If the issue is economic development, what institutions are used for production and trade? If the issue is health, what institutions maintain health – and what institutions cure disease? How well do these institutions currently work to deliver the outcomes that people want? And if they are not working well, why aren't they?

In recent years development professionals and their organisations have become increasingly interested in institutions. Countless development programs have sprung up around the world with names like *institutional development* or *institutional strengthening*. These kinds of programs recognise that institutions matter when it comes to creating positive change.

Institutional development or strengthening programs seek to make the institutions in a particular place work better. They may focus on legal, educational, economic, community, financial or political institutions – or some combination of these. Development programs that focus on improving institutions recognise that institutions can be an important vehicle to deliver better development outcomes.

The problem with mainstream institutional development and strengthening programs, however, is that they tend to start from a predetermined idea of what institutions should look like. Ideas about good 'educational institutions' or effective 'legal institutions' are nearly always imported from outside the local context. These ideas about institutions are based upon how things work elsewhere – they are based upon the institutions of other people, in very different places. When local institutions do not look like outsiders expect, they tend to be overlooked, or judged deficient because they do not meet external criteria for 'good' institutions.

Further, institutional development and strengthening programs nearly all focus at the macro level, with national structures and systems, while generally overlooking the institutions that are already present at the local level. Michael Cernea has argued that in development work, 'institution building at the grass-roots is chronically neglected' as are the 'critical linkages' between micro-level considerations and macro concerns.[46] Thus, for instance, national educational institutions may develop best practice guidelines and curriculum, yet the way local schools work can mean they are ill-equipped to deliver them.

Programs designed to develop or strengthen institutions seldom start with an understanding of the local institutions that are already present on the ground. Yet there are dangers in seeking to create institutional change without understanding the current institutional context. On the one hand, functioning institutions can be undermined and even destroyed, and replaced by less-functional alternatives that are ill-suited to the local context. On the other, large resources can be invested in change that never eventuates – because people will resist institutional change that does not make sense for them.

Rather than starting from a generic, imported idea of what 'good institutions' should look like, anthropologists encourage starting from an understanding of what is already present on the ground. Every institutional context needs to be understood on its own terms. What institutions are already being employed to maintain law and order, educate the young, or enable people to have a voice in decision making?

In many cases, functioning institutions are already present, though they may not fit other people's ideas about what policing, schooling or democratic elections 'ought' to look like. Anthropologists recognise that people around the world make different institutional choices, and that 'non-Western people have always had functional, comprehensible social institutions' with a 'creative capacity to adapt and change'.[47] While institutions may not look like outsiders expect, many institutions function well in their particular contexts – or at least better than the alternatives.

In other cases, however, institutions may be simply lacking, or functioning poorly. They may not be very efficient or very equitable, or they may regularly be placing certain groups at a disadvantage. There are generally reasons why. Understanding why current institutions do not work well is the starting point in the quest for stronger, better institutions.

Current institutions may not work well because they are based on old ideas that have outlived their usefulness. They may have been designed to protect the interests of powerful groups and to exclude others from accessing resources. Institutions may have been previously healthy, but have been undermined or destroyed as a result of war, ethnic conflict or forced migration. In the wake of significant social disruption, groups of people may be left without a common set of structures, rules and norms that they can mobilise to get things done.

Institutions are important in development work, because they facilitate or constrain people's actions, directly affecting development actors' 'room to manoeuvre'. There are many contexts around the world where institutions do not work well, or where they have been designed to privilege the interests of some groups over others. Some groups of people over generations have consistently found themselves in institutional contexts that limit their options: institutions that deny them rights to education or land, for instance, or that normalise unjust treatment because of where they are from, the community they belong to, the language they speak or other characteristics. For others institutional contexts have been more favourable.

Institutional change is always possible, but it is generally neither quick nor easy. This is because people's ways of organising themselves tend to be closely related to how they see and think. People's cultural frames make certain ideas obvious and unquestioned, and other ideas hard to see. It is difficult – though not impossible – for people to shift their cultural frames and think about working differently; to imagine, for instance, that a company could be owned by workers rather than bosses, or that a country could be run by a woman instead of a man. Because people do not change how they think overnight, institutions take time to change too.

Anthropologists call attention to the intimate, up-close workings of how institutional change actually happens. They show that change is a deeply social process, as people communicate, learn from one another, and shift their thinking about what are and are not accepted ways of working. As anthropologist Robert Netting describes it, institutional change happens through the observable processes of 'making of rules and monitoring of behaviour, that goes on in face-to-face corporate groups from the household through the work group or village to the microenterprise and the government bureau'. [48] Institutions guide people; but people also shape and re-shape institutions.

In many places around the world people have developed clever institutional arrangements for organising the kinds of activities that development professionals regularly seek to support and improve. Some of these institutions, such as community-based resource management systems, or indigenous educational institutions, may represent quite innovative ways of delivering development goods. Institutional diversity is a great resource for the future. However, it is a resource that can be easily missed if developers start with predetermined ideas

---

Box 1.5: Case study, *Vivir Bien*

In Bolivia the national government has been reinventing some of its key institutions, attempting to move away from Western ways of doing development to adopt Indigenous Andean ideas and ways of working instead. The policy framework of *Vivir Bien* is an institutional framework based in Andean concepts of *well being* and a *plural economy*. Well being (literally *vivir bien*) is a holistic concept of positive change that moves beyond economic or social targets to integrate cultural and spiritual values. The *plural economy* is a concept of 'economy' that moves beyond capitalism to include community-based economic activities. *Vivir Bien* represents a new way of working at the national level; it is an institutional framework that places Indigenous values and worldviews at its core. Further, it has an explicit aim to benefit marginalised groups. At the same time, this new way of working co-exists with older institutional structures. Lorenzo Soliz Tito (2011), writing on *Vivir Bien*, observes that the values, indicators and ideas of traditional development models still dominate the Bolivian landscape, and the alternative models have been difficult to implement. Ideas have long lives. Institutional change is a slow process, and marginalised rural and Indigenous groups are not starting the process on a level playing field. The desire for change is real, but old ways of working persist: even when things change, they still stay the same.

---

about institutions and what they should look like, rather than appreciating what is already there on the ground.

Institutions across cultures

Anthropology provides an up-close view of how institutions work all around the world. Because institutions work differently in different places, there are challenges when different organisations and communities try to work together. Development actors from different cultural backgrounds often find that they use different structures to get things done, and have different rules and norms about how to act. One group wants a quick meeting; the other wants a deliberative process, for instance.

Anthropologists have often studied these kinds of cross-cultural encounters in development practice. First and foremost, they have been interested in understanding the institutions of the people who are on the receiving end of development work: the 'target groups' or developees. Anthropologists have worked to make the institutions of these groups visible and comprehensible to outsider developers. In recent years anthropologists have also become interested in the institutions of developers as well.

Developers and developees have very different ways of working: different structures, different rules and different norms. Anthropologists have started to look at the meeting points between the two. They ask: What happens when very different groups of people, with very different institutions, come together in a development project, program or partnership? This meeting point or encounter is sometimes called a *development interface* or a *development arena*.[49] The development interface or arena is where different actors from different backgrounds come together, interact with one another and negotiate what 'development' will look like.

This interface is often the site of confusion or misunderstandings. This is particularly the case when developers and developees are from very different countries and very different cultural backgrounds – as is the case in much international development work. Yet confusion and misunderstanding at the interface can also arise when developers and developees are from different regions of the same country, different social classes, or even when they are people who undertake different types of work. Government officials and small business people, for instance, may demonstrate quite marked cultural differences in how they work, as may rural and urban decision makers.[50]

There are numerous examples of how the meeting point of different institutions can cause difficulties, even among development actors with apparently similar cultural backgrounds. In Spain, for instance, anthropological research on a government-sponsored dam project on the River Esera showed that a community advocacy group and the national government had such incompatible ways of working that they could not even communicate effectively.[51] All were Spanish, but the institutions they used to get things done were vastly different; in the end, the community group's arguments simply made no sense within the institutional ways of working of the government.

When developers and developees have very different ways of working, what happens? Often a *cultural brokerage* role emerges, in which a particular person or group of people navigates across institutional contexts to facilitate the process of working together. A number of anthropologists over the years have studied the role of cultural brokers who mediate the relationships between different organisations and communities. A key skill of cultural brokers is that they understand the 'institutional turf' of very different organisations: local communities and government ministries, for instance. Because they understand how very different institutions work, they are in a key position to enable relationship-building.

In the Spanish dam case cited previously, the community group found that in order to dialogue with the government about their opposition to the dam, they had no option but to learn how to work within the government's institutional spaces. They did not know how to do this, however. In the end, anthropologists themselves played a brokerage role: they helped the group recognise that in the government institutions, 'only economic, technical and legal arguments were accepted' and that their own arguments about community survival were not considered valid. Then anthropologists helped community members to learn to 'use the enemy's arguments and to do so in the enemy's terrain'.[52]

This process of learning how to navigate the institutions of powerful groups can be effective. At the same time, it is often symptomatic of deep power inequities. Anthropologists who have studied the interactions between developers and developees have concluded that at the interface, the institutions of developers nearly always dominate relationships. The ways of working of the more powerful group – generally, the developer, not the developee – set the tone of the encounter and become the standard structures, rules and norms in force. In order to work with developers, developees must adapt to their institutional turf.

Thus, studies of participatory development have shown that participatory processes nearly always take place within the institutional 'spaces' of the government or development organisation, rather than in the spaces that developees create for themselves.[53] The only way for people to participate in most development processes is to enter what is essentially a foreign institutional turf: formal office meetings and structured workshops, rather than informal community gatherings or kitchen-table chats. One of the key questions for participatory processes thus becomes: *On whose institutional turf does the encounter take place?*

When developers' institutions dominate development encounters, there are practical consequences. Disadvantaged and marginalised groups are further disadvantaged and marginalised: it is they, not the professionals, who must learn a new set of rules and work out how to function effectively in someone else's institutional space. Not everyone can do this; those who can't are left out.

At the same time functional local institutions are rendered invisible. When developers' institutions dominate the interface, local institutions are de-legitimised. Often development efforts duplicate local ways of working, but in less efficient or adaptive ways. Resources and attention directed to new institutions and 'better' ways of working promoted by outsiders can diminish the legitimacy and sustainability of functional indigenous institutions – and leave locals worse off.

In the end, attention to institutions reveals that culture matters deeply in development work. Different groups of people work differently: there is no single correct way to organise production, trade, education, health care, governance or any of the myriad of human activities that development work engages with. Anthropology demonstrates the diversity and adaptability of institutions on the ground in different contexts. These diverse institutional options around the world provide a rich and often-overlooked resource for positive change.

Summary: Understanding development

This chapter has explored a number of key ideas and insights from the anthropology of development that can directly inform how we think about and 'do' development. Anthropologists have used ethnographic research to study development processes 'on the ground' and 'from within'. Their work with communities and organisations around the world – from the most marginalised, to the mainstream and powerful – tells us a number of important things about development.

This chapter has provided a first look at what *development* looks like when seen through an anthropological lens. The key message of this chapter is that development is a social and cultural process. It is framed by the ways of knowing, thinking and acting that are shared by members of a social group. Development is not a technical process than can be designed and implemented in isolation from people. Rather, development is created in the interactions among people in particular contexts.

Development involves *actors* – people and their organisations – who have different social positions and roles. Development actors, from the powerful to the powerless, are all diverse, savvy and situated in a wide range of contexts. Developers, developees and others all have different characteristics and resources. They all have their own goals, which they pursue as best they can within the room to manoeuvre available to them. Because all development involves people, it is best understood as a social and cultural process of negotiation rather than a technocratic, managerial process of implementation.

Development actors not only have different social positions, they also have different *knowledges* – they know different things, and have different guiding ideas and logics. People around the world have different ideas and logics about what constitutes positive change: there is not one 'development' or one way of achieving it. People also have in-depth knowledge of their own local contexts – situated, indigenous knowledge – which outsiders may lack. Anthropologists have worked to make indigenous knowledges more visible and valued in development work, and have warned of the consequences of development ignorance.

Finally, development actors have different ways of working – that is, different *institutions*. Institutions are the structures, rules and norms shared within a social group that define how things are done. Institutions directly affect the options and room to manoeuvre that development actors have. Around the world, people organise themselves in a wide range of ways to conduct economic and social activities. Yet developers' own institutions have tended to dominate development encounters. This has placed other development actors at a disadvantage, and meant that their own diverse institutional resources have been overlooked.

This chapter has presented a set of ideas distilled from anthropologists' up-close observation of and engagement with development. Anthropologists' work with marginalised groups, their work with developers, and their in-depth observations of the interactions between and among different development actors, reveal that development is a social and cultural process. Different development actors pursue their goals from different social positions, within different cultural frameworks and with differing amounts of room to manoeuvre.

There are many other insights anthropology can provide to development professionals about specific aspects of social and economic change. But the ideas in this chapter are a good place to start. These ideas come up again and again when anthropologists write about development. And each is important at a practical level for those who want to do development work that makes a positive difference.

Notes

1 This is the definition of sustainable development used in the Brundtland Report (United Nations 1987).
2 UN (2015).
3 See Andrea Cornwall and Deborah Eade's collection *Deconstructing Development Discourse Buzzwords and Fuzzwords* (2010); the quote is from p.viii.
4 Hopper (2012, p.54).
5 Ortner (1984, p.143).
6 Harrison (2015, p.155).
7 Chambers (1983).

8   The term 'applied anthropology' refers to the use of anthropology in practical contexts. See, for instance, the Society for Applied Anthropology: www.sfaa.net.

9   Standing (2013) notes that economics remains the 'leading authoritative discipline' in development theory and practice, while anthropology is 'often relegated to a minor status'.

10 Green (2012, p.53).

11 Standing (2013, p.5).

12 Eyben (2014, p.17) uses the small-d/ big-D distinction, which she attributes to Gillian Hart (2001). It is important to note that some of these researchers prefer to call themselves economic anthropologists or political anthropologists, rather than anthropologists of development.

13 Eyben (2014).

14 The language of 'ideas and practices' in development is borrowed from Mizanuddin (2013).

15 Simone Abram and Jacqueline Waldren's (1998) collection on local development is one exception; they note that the relationship between anthropology and development 'has largely been dominated by a focus on international development'. Other exceptions include Lisa Peattie's work in urban planning in the US, Elizabeth Harrison's work on community development in the UK, and Robyn Eversole's work on regional development in Australia.

16 There is a longstanding distinction in anthropology literature between 'anthropologists of development' and 'development anthropologists' (see, for instance, Hoben 1982). The former study development; the latter practise it. This distinction is misleading, however, as many anthropologists these days do both.

17 Some examples include Grillo and Stirrat's (1997) *Discourses of Development* and Cornwall and Eade's (2010) *Buzzwords and Fuzzwords*.

18 In sociological terms, development actors have *agency*. Sociologists like Norman Long have thus explicitly named 'actor-oriented' development approaches (e.g. Long 2001) as part of a theoretical positioning that focuses on the agency of actors rather than, by contrast, big 'structures' like capitalism or globalisation. Anthropologists, while certainly not uninterested in big structures, have always tended to start from an on-the-ground assumption that actors are important, without explicitly naming their approach as 'actor-oriented'.

19 Olivier de Sardan (2005) uses the term 'room to manoeuvre' as a convenient way of expressing the tensions between actors who act and the situational limitations on their ability to do so.

20 Anthropologist Jean-Pierre Olivier de Sardan (2005) uses this terminology of '*developers*' and '*developees*' – those who instigate formal development initiatives and those who are on the receiving end of them.

21 This is referred to as the 'Aidland' literature; see Harrison (2013) for various references and a critique.

22 Tendler (1975).

23 Mosse (2005).

24 Crewe (2014).

25 Uddin (2013).

26 Larsen (1998, pp.26, 31).

27 Talle (1998, p.37).

28 Escobar (1995), republished in 2012.

29 Ferguson (1994).

30 Abram (1998, pp.8, 13).

31 Boissevain and Theuma (1998, p.114).

32 Chambers (1983).

33 For instance Chambers *et al.* (1989), Scoones and Thompson (1994) and Scoones *et al.* (2009).

34 For instance Brokensha *et al.* (1980) and Warren *et al.* (1995).
35 For instance Sillitoe *et al.* (2002), Pottier *et al.* (2003) and Bicker *et al.* (2004).
36 Eversole (2015, pp.95–98).
37 Larsen (1998, p.27).
38 Richards (1985).
39 Sillitoe (2015, p.13).
40 Hobart (1993).
41 Olivier de Sardan (2005).
42 Ibid.
43 North (1990).
44 Acheson (1988).
45 Netting (1991, pp.148–149).
46 Cernea (1986, pp.xiii, xv).
47 Netting (1991, p.151).
48 Ibid.
49 The term 'interface' is particularly associated with the work of Norman Long, a development sociologist. The term 'development arena' is used by the anthropologist Jean-Pierre Olivier de Sardan (2005).
50 In Australia anthropological research on regional development has shown how on-the-ground institutions in rural communities work very differently than urban decision makers expect (Eversole 2016).
51 Mairal Buil and Bergua (1998).
52 Ibid.
53 See Cornwall (2008), Eversole (2012).

2

Anthropology of development in practice

Chapter 1 introduced a number of core ideas from the anthropology of development which are directly relevant to development practice. It described how anthropologists seek to understand development situations on the ground and from within, and what their perspectives reveal about how development works.

Chapter 2 takes these insights from theory and illustrates what they look like in practice: on the ground in real places. This chapter shows how the ideas about development discussed in Chapter 1 matter for real people around the world. It aims to make the theoretical insights from Chapter 1 come alive through real-world stories.

Using anthropologists' in-depth accounts, Chapter 2 shows how paying attention to actors, knowledges and institutions can explain a great deal about how development actually happens. Core ideas from anthropology can explain why apparently good, well-designed programs fail, and why projects and policies can end up looking very different than planned. Further, they reveal unsuspected on-the-ground resources for positive change, and what the empowerment of marginalised communities looks like in practice.

The first part of this chapter focuses on 'small-d' development: social and economic change in the absence of formal, planned development initiatives. This section presents a set of case studies of how diverse local communities around the world confront and create change for themselves – and what development professionals can learn from watching them. These stories reveal untapped potential for development professionals to engage with a range of development actors who are already championing change.

The second part of the chapter explores 'big-D' Development – planned development initiatives and interventions by governments, bilateral organisations and NGOs. The case studies in this section show what can be learned from an up-close look at formal development projects and programs, across a range of country contexts and scales. These case studies encourage development professionals to reflect on their own roles as development actors in context, and discover ways that their work can challenge – rather than reinforce – the relationships that perpetuate poverty.

Stories of change on the ground

Many anthropologists have studied social and economic change processes. They have often observed change happening even when it was not intentionally planned. Yet these changes have often been profound: new ideologies, new lifestyles, new ways of organising work, households, settlements and so forth. History is full of examples of change – and most were not the result of intentional planning.

Development work aims to intentionally create positive change. But effective change-makers recognise that planned change is not the whole story. Their work takes place on a broader landscape. For development professionals, projects and programs may be front and centre in importance in their minds. But larger change processes are already at play in every community and every context where development professionals work. Change never takes place on a blank slate.

For development professionals, understanding how small-d development works is key to effective big-D development practice. The places where we work have histories. Each place contains local actors who helped to make things as they are, and who continue to act in the present. Understanding the change that has happened – or failed to happen – in the past can inform current change efforts. Equally, understanding what actors are present, and what they are currently doing, can reveal a range of unexpected resources and opportunities.

Case studies of small-d development illustrate how successful social and economic change can come from unexpected places – with unexpected people in the driving seat. For development professionals, the stories in this section show the importance of paying attention to grassroots actors: who they are, what they know and how they work.

Grassroots actors

The writings of anthropologists are rich with stories of real people in real places. Anthropologists have a talent for making far-away people come alive on the page. Whether these are Nuer or Xavante – or North American lobster fishermen – the people described by anthropologists feel real. Stereotypes about what far-away people are like, and what they can and can't do, fall away. Their ways of thinking and ways of working, even if they seem very odd and foreign on first look, start to make sense. It is possible to feel as though we, like the anthropologists, are seeing these groups up-close.

From a distance, it is easy to stereotype particular groups of people. It is easy to assume that they are passive and unchanging – or the victims of change instigated by others. Indigenous and colonised peoples, for instance, are often portrayed as 'traditional', as if they were stuck in the past. Smallholder farmers and artisans are described as 'poor', 'peasants' or leftovers from a 'pre-industrial' economy. And women are often assumed to be more concerned with maintaining households and raising children than driving economic and social change.

Studies by anthropologists have actively challenged these kinds of stereotypes and assumptions. Their up-close view has shown that groups considered passive or marginalised are often in fact playing an active role in change processes. Anthropologists have documented 'local forms of development' and 'internally generated change' in rural and marginalised settings all around the world.[1] The case studies that follow show some of the many ways that change can be driven from the grassroots and created by people on the margins.

Case study one: Actors without history?

European historians have long portrayed the arrival of Europeans as the beginning of history for colonised peoples. Anthropologists, however, have observed that these societies have their own histories, of which colonisation is only one part. Eric Wolf's influential book *Europe and the People without History* describes how so-called 'traditional' peoples around the world have always changed: interacting with each other and with colonial powers through complex processes of trade, expansion, war, conquest, alliance-building and other interactions.

Wolf describes 'the world in 1400' – before European expansion overseas – as highly interconnected and dynamic. Around the world, different groups were coming into contact, renegotiating old boundaries and establishing new political, social and economic relationships. In the Andes region of South America, the Inca were one small group among many before they rose to dominate an empire.[2] The subcontinent that is now India had waves of foreign conquerors until the Mughal dynasty established under Babur in the 1500s – yet another group of conquerors – came to be characterised as 'traditional India'.[3] In these and other stories, *Europe and the People without History* questions the stereotype that 'traditional cultures' were isolated, static and unchanging.

Further, Wolf shows, over and over, how groups of people on the periphery of European expansion were always active participants in historical change processes. He takes an up-close look at global processes – such as European conquest, the fur trade and the slave trade – to show how they

continued

played out in particular places, with local actors in the driving seat. An up-close look at the colonisation of northern North America, for instance, revealed complex and dynamic interactions among trappers and fur traders that dramatically transformed economic and social relationships in Native American communities. New commodities and technologies became available, and people changed how they organised themselves to access these new resources. The politics of resource access played out in a complex setting of rival tribes and differently situated actors. Actors of all kinds responded to 'big-picture' change processes, and instigated change in their own right.

Development projects have often taken a similar approach to the old European historians: forgetting what has come before, and starting history from now. This case study reminds us that development work enters landscapes with long histories, where actors are already present who have histories of navigating change.

Case study two: Local economic actors

Development often aims to make local economies work better: to be more productive, more equitable or both. Yet the economies that are the focus of these change initiatives have their own histories, and their own key actors. Anthropologists have studied economic change around the world, and their work can give us important insights about how local economies work on the ground.

Clifford Geertz studied social development and economic change in Indonesia in the 1950s. His book *Peddlers and Princes* compared the very different economic development trajectories of two Indonesian towns: one (Modjokuto) a Javanese 'market town', and one (Tabanan) a Balinese 'court town'. The book focuses on grassroots economic actors in these two towns: on the one hand, the 'entrepreneurial group of Islamic small businessmen' operating in the bazaar-style economy of the Javanese town; on the other, the 'nascent entrepreneurial class of displaced aristocrats' working in the context of traditional agricultural villages.[4] These two different groups of actors – whom Geertz describes in shorthand as 'peddlers' and 'princes' – each played an active role in driving economic development in their respective towns.

Both the 'peddlers' in the Javanese town and the 'princes' in the Balinese town were interested in developing successful firms. Both had clear goals, and both wanted to reap benefits from emerging economic opportunities in the national economy. Both worked proactively to do so – they were, essentially, savvy development actors. But it is there that the similarities ended. 'Peddlers' and 'princes' had very different ambitions. The peddlers sought economic wealth, while the princes sought political power.

Furthermore, each group was situated in a very different social and institutional context. As a result, they faced very different constraints to building successful firms. In Java, the highly competitive and individualistic nature of the bazaar economy limited local entrepreneurs' ability to organise larger firms or raise capital; while in Bali, the embeddedness of economic activities in collective social structures and patron–client relationships meant that firms grew large and were well capitalised, but were often inefficient. There were direct connections between economic activities and their social context.

Peddlers and Princes is an engaging account of economic change playing out on the ground in local contexts. It shows how local people were key actors in larger economic development processes. In a part of the world that has often been on the receiving end of development assistance from elsewhere, Geertz presents a view of savvy local people actively pursuing change in their own right.

This case study reminds development professionals to pay attention to local actors and what they do. These people and organisations are likely to be surprisingly savvy – but also situated in particular contexts that directly affect their room to manoeuvre, and thus their ability to drive change.

Case study three: Women as economic actors

Gender is a social category that has real implications for development actors' room to manoeuvre. A person's social position as a man or woman in particular contexts will directly influence his or her ability to own land or other property, access resources and labour, gain skills, access credit, make household economic decisions or otherwise manoeuvre as an economic actor. Further, work itself is gendered: in a given social context, women are often expected to do certain kinds of activities, and men to do others.

When it comes to the gendered nature of work, roles vary from place to place: there are not universal categories of 'women's work' or 'men's work'. For instance, many agricultural development programs designed by Europeans and Americans have assumed that men are always in charge of farm work. 'Farmers' are assumed to be men. This assumption has been based

continued

on certain culturally framed ideas about gender roles. Yet an up-close look at agricultural economies shows that in many places women are the farmers; or men and women farm together, taking on different kinds of roles.

Sarah Hamilton's *The Two-Headed Household* looked at gender roles in Ecuadorian farming households, where women and men are both involved in farm work. The rural households that Hamilton studied did not have a single male 'head', as many outsiders assumed. Rather, men and women organised their work and economic decisions collaboratively. These collaborative economic practices continued in these rural households over time, even in the face of strong patriarchal influences from the larger society. These included the frequent tendency for agricultural development programs to assume that the 'farmers' were men, and to manage their programs accordingly.[5]

Gender roles in different contexts may be highly rigid; or they may be flexible and negotiable. Importantly, gender roles may change over time, and these changes may give women more or less access to resources and influence. Melissa Leach's work with the Mende of Sierra Leone in the late 1980s documented how their move from upland to lowland rice farming had significant impacts on men's and women's economic roles. In upland rice farming, women and men worked together, with men clearing the land and women planting it. As households moved into lowland rice farming, however, women took on more responsibility for rice, while men started to farm coffee and cacao. The new gender roles significantly changed the economic and social dynamics of these households.[6]

These case studies remind development professionals that actors are diverse and social roles matter. Gender roles have direct bearing on the options that people have and the kinds of resources they can access. These roles do not, however, necessarily look the way we might expect when thinking about women's or men's roles through our own cultural frames.

Invisible knowledges

Development workers often assume that the knowledge that is needed to create change must be brought in from elsewhere to fix deficits in poor areas. Ideas like *technology transfer* and *expert assistance* are used liberally in wealthy and poor countries alike, offering to provide local people with the knowledge they lack. Anthropologists, by contrast, tend to pay attention to what development actors at the grassroots know, and how they use this knowledge to drive change.

Anthropologists have documented how people – even those who are highly disadvantaged – have in-depth knowledge about their context and mobilise this knowledge to manage change and develop solutions to local problems. This does not mean that external technology and expertise are not wanted or needed. But external assistance never enters a landscape void of knowledge.

The following case studies give some practical examples of indigenous knowledges – technical, experiential and cultural – and some of the ways that grassroots actors use their knowledge to improve their circumstances. Further, these stories illustrate that actors at the grassroots generally have sound notional and strategic logics about change. Nevertheless, indigenous knowledges and logics often remain invisible to outsiders. There are, however, real, practical consequences when outsiders fail to recognise them, or when they assume that marginalised groups are incapable of managing change.

Case study four: Knowledge for managing natural resources

Many anthropological studies have focused on the role of local knowledges in managing natural resources in particular local contexts. They have documented local practices of soil management, water management, erosion management and pest and disease control, for instance, in a range of contexts around the world.

The 1995 collection *The Cultural Dimension of Development* brings together numerous examples of resource-management strategies actively employed by local groups in different contexts around the world.[7] These include:

1 Clever agroforestry techniques, such as those used by Indigenous hill communities in Nepal: farmers plant a shrub that grazing animals do not eat as a companion plant to ensure the successful propagation of an important fodder species;[8]

2 Productivity-enhancing farming practices, such as the *chinampa* raised field system used in Mexico and Central America: raised fields allow a wide range of crops to be grown intensively in wetland settings and with low incidence of plant diseases;[9]

3 Indigenous soil and water conservation methods, such as biological and mechanical practices used by male and female farmers in Mali and other regions of sub-Saharan Africa, shown to be effective even though they were largely disregarded by soil conservation experts from elsewhere;[10] and

4 Social and institutional systems for managing common-property resources, such as the social controls established by pastoralists to manage ranges and communal wells in arid and semi-arid Africa.[11]

continued

These examples demonstrate some of the many ways in which grassroots actors around the world have detailed and in-depth knowledge of how to manage resources creatively and successfully in their local environments. Often, these forms of knowledge have been developed over time through experimentation. Nevertheless, outside decision-makers have often failed to notice these local strategies, or have judged them inadequate or unscientific. These case studies remind development professionals to pay attention to the on-the-ground knowledge that already exists and its value for achieving development goals.

Case study five: Knowledge for coping with adversity

Local people often have their own strategies for coping with adversity. These strategies are based on cultural, technical and experiential knowledge about 'how things work' in the local context, and notional and strategic logics about the options for change that are available to them. While local strategies to cope with adversity may not be complete or ideal, failure to understand them can mean that efforts to help can do more harm than good.

Natural disasters such as floods are one example. In many contexts, local people have established their own mechanisms for coping with regular flooding. Rosalind Shaw has observed that in Bangladesh floods can be either hazards and/or resources depending upon one's social positioning. Floods can bring destruction, but also benefits such as improved soil fertility. Local actors have developed their own knowledge and strategies for dealing with flooding, but are less confident in dealing with the flood mitigation strategies of external agencies like the World Bank; while promising to reduce flooding, these also cut off access to key resources.[12]

Local strategies for coping with adversity are often invisible to outsiders, and so efforts to create change proceed with no understanding of the logics and strategies that are already in place. In a recent ethnography of street children in Mexico, Norman Long has observed that attempts by police and social workers to improve the circumstances of street children were ineffective, precisely because they were designed by people who had no experiential knowledge of 'the everyday experiences and predicaments of street living'.[13] Long argued that 'the children's own knowledge and experience' was an unrecognised resource; the well-meaning adults who designed the program had no understanding of the strategies that street children themselves had developed to manage their situations. This created disconnects between the children and resource-rich adult institutions – and ultimately led to more, rather than fewer, children on the street.[14]

These case studies illustrate the importance of recognising that even marginalised and disadvantaged groups have important knowledge about their situations and logics about how best to cope in difficult circumstances. Failing to recognise this risks marginalising them further by replacing local knowledge with development ignorance.

Case study six: Knowledge for innovation

Anthropologists' up-close view of local change processes suggests that *innovations* – new and improved ways of doing things – often emerge on the ground as the result of experimentation and adaptation. In some cases local knowledge and knowledge from elsewhere are combined. In other cases local knowledge is simply mobilised in new ways in response to changing conditions.

Anthropologists have written extensively on how indigenous knowledge is 'practised' and 'performed'. Rather than a static body of knowledge, local knowledge often manifests in how people act in response to continually changing circumstances. Writing on farming in central Africa, James Fairhead has observed that:

> In Bwisha, the dramatic and ceaseless changes in local agriculture and social organization over the last seventy years mean that each new generation of farmers faces new agricultural problems. Local knowledge is better envisaged as empirical and hypothetical. . . . It is living and dynamic.[15]

Paul Richards has similarly suggested that indigenous knowledge can be understood as 'a set of improvisational capacities called forth by the needs of the moment'.[16] The results of this 'improvisation in action' are eminently practical: for instance, trying out different crop varieties has built up deep local knowledge of which cultivars work best in particular environments, and the best methods for growing them.[17]

Innovations thus may literally emerge from the grassroots, as people practise and perform their knowledge in response to new and emerging circumstances. For instance, in Kerala, India, artisanal fishermen developed

continued

artificial reefs to replenish the fisheries resource that had been decimated by mechanised trawling.[18] In Nigeria, 'countless small-scale artisans, traders, and farmers' developed a national cassava industry in the 1980s and 1990s to replace dependence on cereal imports. They developed cassava-processing machines, the most successful of which were made by small-scale artisan producers using locally available materials in response to customer needs.[19]

These case studies remind development professionals that the world is full of change makers and innovators, even in places that are poor or far from centres of power. Paying attention to development actors and what they know can reveal a range of unexpected ideas and resources for change.

Institutions and power

Anthropologists are often concerned to point out the ways that one social group may exercise power over another: colonisers over colonised, for instance, or wealthy over poor. Anthropologists have looked up-close at how some groups are able to limit other groups' room to manoeuvre regularly and predictably: by blocking access to resources, for instance, or by depriving them of voice and influence. When these kinds of power relations are examined up-close, a common theme is that less powerful groups are often forced to work within the institutional rules of dominant groups.

Institutional change is a frequent focus for development work. Yet most developers have little understanding of people's on-the-ground institutions or how they have emerged over time. Anthropologists' attention to institutions – 'the way things work' in particular social settings – reveals a great deal about what is needed to create real social and economic change.

Anthropologists pay attention to the different institutions that people use to organise their households and communities, to allocate work, to govern resource access, to maintain social order and to manage risk in very different settings around the world. They explore what these arrangements mean for different kinds of development actors: for women and men, foreigners and locals, or land-owners and labourers, and how these institutions may expand or curtail their room to manoeuvre. The following case studies provide some examples of institutions in practice.

Case study seven: Institutions for managing at the margins

Some communities seem to be desperately in need of help. They are living in adverse circumstances – homeless or in poor housing; poorly clothed and fed; working in demeaning activities. Yet a closer look shows that their activities, even in highly adverse conditions, are seldom desperate. Rather, even in difficult conditions, groups of people organise themselves to manage their situations, and develop institutions to help them do so.

Anthropologists have documented functional institutions in a range of informal and unlikely places. Robert Netting has observed that 'Officials may not believe that the villagers have their own regulation to limit firewood cutting or that urban ragpickers organize their scavenging' – but anthropologists have documented these practices in detail.[20]

Street children, slum dwellers, any social group – however marginalised from 'mainstream society' – generally have their own structures, rules and norms in force, which frame their daily activities and enable people to organise themselves and cope even in very challenging circumstances. Because anthropologists often work up-close with marginalised groups, their work often reveals otherwise invisible institutional arrangements – the institutions that people at the margins use to survive.

Ignoring these institutions is a typical recipe for development failure. For instance, Michael Horowitz has demonstrated how a series of livestock development projects in Africa nearly all failed to achieve their aims: 'Degradation was not reversed; productivity was not increased; and herder incomes, rather than rising, generally fell.' Herders were working in a marginal environment, and so to manage that they were already mobilising a range of complex social arrangements and practices to regulate access to scarce pasture and water. Horowitz observed that the projects to help herders all failed because 'they fundamentally misunderstood the ecology and sociology of pastoral production systems' and the alternatives they proposed did not deliver on basic livelihood needs.[21]

Attention to local institutions can communicate new respect for development actors at the margins and their ability to manage situations where resources are regularly out of reach. In *Between Field and Cooking Pot*, for instance, Florence Babb spent time with market women who work in the 'informal economy' in Peru. Her research revealed a range of savvy economic and political strategies that these market women were employing to grapple with challenging and rapidly changing economic conditions.[22]

For development professionals, being aware of the institutions that people already use to manage at the margins means that change efforts recognise and build on, rather than undermine, strategies and practices that people have developed for themselves to cope in difficult circumstances.

Case study eight: Institutions and Indigenous disadvantage

Indigenous groups are often among the most economically and socially disadvantaged groups in the countries where they live.[23] This is unsurprising when we consider that, for generations, Indigenous groups have had to navigate and work within the institutional environments of their colonisers. Indigenous ways of working have been regularly subsumed under more powerful colonial and post-colonial institutions, which have tended to regularly deprive Indigenous peoples of both resources and legitimacy.

The roots of contemporary Indigenous disadvantage can be traced back historically to conquest and colonisation: outsiders moved into local contexts and re-wrote the institutional rules. New economic, social and political arrangements were established. Land, now enclosed or cultivated, became unavailable for traditional activities such as hunting and gathering or migratory pastoralism. New rules of ownership and exclusion came into force, and longstanding community livelihood practices were made illegal, or became impossible to maintain. With colonisation, accepted ways of doing things changed very quickly. Functional Indigenous institutions were undermined and sometimes completely destroyed.

In contemporary societies, there is an increasing awareness of Indigenous peoples, and a growing commitment to Indigenous rights. Yet in practice, disadvantageous institutional arrangements continue. The dominant institutions in post-colonial societies are nearly all non-Indigenous. Schools, health care providers, legal systems and so forth are all based on non-Indigenous ways of working. Thus, Indigenous groups continue to navigate foreign institutional turf on a daily basis.

In Australia, for instance, the government has recognised that Aboriginal Australians have a right to claim ownership of their traditional lands – a process called Native Title. Yet claiming Native Title requires Aboriginal groups to navigate the complexities of the Australian legal system, which is a Western institution. In this institutional context, claims about what qualifies as 'traditional' are defined in Western, not Indigenous, ways.[24] Similarly, while the government has processes in place to protect Aboriginal heritage, 'heritage' is defined in terms of Western nature/culture binaries that make little sense from within Indigenous worldviews.[25] In Australia, a number of anthropologists work as cultural brokers to help Aboriginal communities navigate the legal system and put their cases for Native Title and heritage recognition in a language that makes sense within Western legal institutions.

Indigenous ways of working have changed and adapted to colonial and post-colonial contexts, but they still differ from those of mainstream 'Western' institutions. Many Indigenous institutions do similar things to

Western institutions – they provide education, social safety nets or governance, for instance; but they do them in different ways. For instance, Indigenous Hawaiian educators have worked to establish community schools in Hawaii, based explicitly on Indigenous Hawaiian pedagogy that is culturally and family oriented and place-based.[26]

For development professionals, understanding that Indigenous communities often have different ways of working than the dominant societies in which they live is an important first step to developing more respectful working relationships with Indigenous groups. Internationally, the recognition and practice of Indigenous institutions under the banner of *self-determination* are increasingly recognised as key to combatting Indigenous disadvantage.

Case study nine: Changing the institutional context

Institutions – from post-colonial states to globalised markets – often favour the interests of certain groups over others. This can severely limit the room to manoeuvre of people who are not advantaged by these arrangements, curtailing their access to resources and influence. Nevertheless, people are seldom passive victims of their institutional environment. They still find ways to challenge and sometimes shift the institutions that disadvantage them.

A number of anthropological studies have looked at local organisations and the roles that they can play in shifting institutions to become more efficient and equitable.[27] Such 'grassroots' organisations are often not visible to outsiders. Nor do they necessarily characterise what they do as *development* work at all. Nevertheless, these organisations are often instrumental in creating new institutional options for local communities that provide alternatives to dominant ways of working.

For instance, some local organisations are established as a strategy to rework market institutions so that they deliver more benefits to local people. In *Weaving a Future*, Elayne Zorn described how the Indigenous residents of Taquile Island in Lake Titicaca developed a successful model of community-controlled tourism based on their income from textile production. This model relied on strong local organisations to drive it. These organisations were able to function as brokers between local communities and tourists, and were able to re-shape the tourism economy in a way that delivered greater benefits to local residents. Though not without problems and conflicts, the Taquile case shows how people at the local level organised themselves to

continued

challenge disadvantageous market institutions and shape new ones on their own terms.[28]

Anthropologists suggest that these organic forms of self-organisation may be superior in practice to external developers' efforts to create institutional change. In *Raising Cane*, for instance, Donald Attwood explores the experiences of grassroots sugar-cane growers' cooperatives in Maharashtra, western India, compared with the experience of government-established sugar cooperatives in the north of the country. Both aimed to make production more efficient and deliver more benefits to local farmers, but the latter were found to be largely corrupt, inefficient and highly bureaucratised, while the former – most founded and led by 'peasants' from village backgrounds – were successful and flourished.[29]

There are many examples of institutional change being led from the grassroots, as actors on the ground experiment with new ways of working that suit their own aspirations and contexts. For development professionals, this means that there are already many potential allies, and many experiences to learn from, in the quest for positive change.

Learning from change on the ground

Development professionals often feel as though they, alone, are charged with creating change. Yet the case studies in this section have shown that social and economic change happens independently of the actions of developers. Further, they show that understanding the nature of change processes on the ground in real contexts is vital to effective development work.

These brief cases from the anthropological literature illustrate some of the ways that development actors at the grassroots create economic and social change, even in the absence of formal development initiatives. Within the room to manoeuvre that they have, peddlers, fur trappers, farmers, local leaders, market women and a range of other social actors pursue their own agendas, often in clever and resilient ways. They mobilise their knowledge about local contexts and use it to develop ways of working that meet their needs. These local knowledges and institutions are usually invisible to outsiders, but are key to how people around the world survive adverse conditions, manage rapid change and work to create the futures they want.

All around the world, grassroots actors make change happen and respond creatively to change thrust upon them from elsewhere. For development professionals, paying explicit attention to local people and organisations, what they know and how they work, can teach us a great deal about change. It can show us, for any given context, what things are possible to improve, and where the obstacles lie. It can help us to see old problems in new ways, understand their deeper causes and reveal on-the-ground knowledge, resources and energy for change.

Stories of development practice

The case studies in the previous section described social and economic change taking place in the absence of formal development programs or projects. This section turns to studies of formal, planned development interventions: development practice itself. The case studies in this section explore what happens when development organisations of various kinds seek to intentionally create change.

Anthropologists' case studies of planned development explore how development organisations, projects and programs work. They unpack the relationships among 'developers', 'developees' and other development actors, providing a detailed and dynamic view of the diverse individuals and organisations that are involved in development processes. Understanding these relationships has practical implications for the effectiveness of development work, especially its effectiveness in challenging the relationships that perpetuate poverty.

Actors in development practice

Anthropological studies of development work look up-close at development initiatives and organisations. They may start with the experiences of the people on the receiving end of development – the diverse 'beneficiaries', 'target groups' or 'local partners'; or they may start with the professionals, staff and volunteers tasked with delivering development outcomes. Whatever the starting point, anthropologists pay attention to *who* is involved in development work – the actors.

Anthropologists recognise that actors in development work are diverse – with different genders, ages and social positions; working in different professional and/or community roles; identifying with different organisations and communities, with different levels of interest or commitment to them. These development actors are savvy, situated in social contexts, and they interact with one another around specific ideas and initiatives to create change. The interactions among development actors are important, because they directly shape the practical outcomes that result.

The case studies that follow reveal that there are many actors involved in development – well beyond those captured in most project management tools. A project proposal often contains a list of partners and what they are contributing; a project logframe will detail inputs, activities, outputs and outcomes, the key people who are responsible for delivering them, and who is expected to benefit. On the ground, however, many other people and organisations affect, and are affected by, development activities. As this larger group of development actors interacts with each other, they create development outcomes, good and bad. As Rosalind Eyben has argued, in the end, *relationships* are central to effective development work.[30]

Case study ten: Invisible local actors

Development professionals often simply overlook local actors: choosing to focus instead on the problem to be solved. Technical solutions are designed and implemented in response to practical development problems. Yet the actors that will have to operationalise these solutions are regularly ignored. And the idea that local actors could have their own systems and solutions already is not even on the radar. The simple invisibility of actors 'on the ground' has created numerous development failures, often leaving local people worse off than before.

Michael Horowitz's work with irrigation projects in Nigeria and Mali in the 1980s demonstrates the dangers of overlooking local actors. Project planners, implementers and funders all assumed that new, irrigation-based farming systems would be superior to local farmers' traditional practices. Focusing narrowly on the crop yields that could be produced by mobilising water for irrigation, developers argued that the new schemes would raise farm productivity and make local people better off. However, as Horowitz observed, they failed to see all the other ways that local actors were already using this water for productive purposes.

Horowitz looked up-close at farming communities and revealed how local actors were mobilising their resources in functional and highly efficient production systems. He calculated that while irrigated agriculture did indeed create higher crop yields as predicted, the overall net returns to labour and capital were significantly *lower* for irrigation than for traditional agriculture. Further, there were unsuspected social costs. Irrigation significantly increased workloads for women – with no corresponding increase in land rights – and removed water from other productive local uses, such as fishing, cattle, and household water access. Horowitz calculated that the rate of return to water was *1500 times better* in the existing local schemes than in the new schemes.[31]

When developers fail to see local actors, they can completely misread the development 'problem'. James Fairhead and Melissa Leach's analysis of environmental policy in Guinea in the 1990s provides a cautionary tale. Distant government policy makers interpreted West Africa's patchy forest–savannah transition zone as 'the last endangered relics of a once extensive natural-forest cover now destroyed by local farming and fire-setting'. They blamed local villagers for destructive environmental behaviour, and mobilised an environmental degradation narrative to attract donor funds to protect the forests.[32] Meanwhile, the anthropologists looked closer at the local villagers to understand what they were actually doing.

They discovered no evidence of destructive locals; rather, villagers over generations had actively been planting forest patches around their settlements. Patchy forest, it turned out, was not a sign of environmental degradation. Rather, as aerial photographs of the area confirmed, the amount of forest area around the villages had actually been increasing over time – not decreasing as policy makers had assumed – due to the active forest stewardship by local people.

The invisibility of local actors in the environmental policy process meant that these valuable forest management practices were misunderstood. Then they were actively suppressed. In the name of protecting an endangered environment, government authorities in Guinea began to control villagers' access to forest resources. They started to criminalise traditional resource management activities. The villagers were left with fewer resources, and less room to manoeuvre.

These case studies illustrate the real problems that can emerge when development workers focus on problems and solutions, but fail to notice local actors. Ignoring local actors creates misunderstandings, missed opportunities and unintentional – but often very real – harm. Yet, even if the developers had chosen to consult with local communities about the development 'problem', the results may well have been similar. When developers make assumptions about the nature of the problem, this can easily render local people invisible and unheard.

Case study eleven: Professionals as development actors

In their quest to solve development problems, development professionals often overlook local actors and their practices. One reason for this is that development professionals have their own cultural frames: their shared ways of seeing, thinking and acting. They work in organisations or professional frameworks that encourage them to focus on certain things, and to de-emphasise or ignore others. Often, these professional frames make it difficult to see other development actors or why they matter.

A vivid example comes from anthropologist Lisa Peattie's work on urban planning. In a case study from the 1960s, she described how the professional frames of planners and urban developers led them to exclude local people and organisations from any involvement in a major Venezuelan urban

continued

development project. Planning professionals saw urban development as largely a question of design: designing an attractive city. Since 'planning was thought of as design, rather than as institution-building or organizing, it seemed entirely reasonable to do it in Caracas, 350 miles away' from the blank-slate site of the planned city – where around 50,000 people already lived.[33]

The results of this framing of urban development can be observed right into the present: the city has turned out to be 'inefficient and unpleasant' with a sharp divide between rich and poor: 'now three-quarters of the population lives in shanty settlements with few or no services, while across the river rise the pricy condominiums of the modern city.'[34] Further, Peattie warns that today's urban planning approaches are not immune to similar problems. New frames that picture 'the city as an economic system' and an arena for economic partnerships between government and for-profit enterprises are equally problematic. By making market logic appear 'as much a given as gravity' and leaving political and power issues outside the frame, these approaches continue to support vested interests and reinforce economic divides.[35]

Even when development professionals actively seek to support the agendas and interests of less powerful actors, their own frames can get in the way of effective change-making. *Rights-based development approaches*, for instance, attempt to shift the focus of development work toward respecting and protecting the rights of less powerful actors. These approaches emphasise that, as human beings and as citizens, people have a *right* to things like food, housing, education, security and political voice. While this is an important shift, it also has some significant limitations. Rights-based development approaches define *rights* according to the Western concept of an abstract, legal individual. This is very different from community-based understandings of rights, where rights are relational, and often include a range of related responsibilities and entitlements. As a result of the way professionals think about rights, actors who work within collective and community-based approaches may struggle to make sense of what these ideas actually mean for them.[36] Professional frames, even with the best of intentions, can render local actors and their perspectives invisible.

These case studies suggest that it is important for development professionals to be conscious of how they 'frame' development issues. Development professionals each have their own personal and professional ways of seeing, thinking and acting; this affects what they see, and what they miss, when they look to create positive change.

Case study twelve: Actors and relationships

Development professionals are diverse, and so are the actors on the receiving end of development initiatives. A number of anthropological studies have looked up-close at the relationships between different kinds of development actors, and what these relationships mean for change-making in practice.

David Mosse has observed that development workers tend to conceive of development projects as 'closed and controllable systems'; yet in practice, projects and programs play out in highly complex social settings.[37] Mosse's study of a participatory agricultural development project in India described the complex relationships among development actors associated with the project: farmers, project staff and community organisers from different backgrounds, project managers, village big-men and other elites, women, government officials, *jankars* (local 'knowledgable people'), consultants and others. He showed how these diverse actors manoeuvred, within the framework of a particular project, to build relationships that would provide them with access to resources and support of various kinds.[38]

Mosse's case study makes the point that project implementation is always a deeply social and political process. Social relationships create project activities and shape project outcomes. This raises the question: What kinds of relationships among development actors can produce effective social change?

Rosalind Eyben and Rosario León provide an insightful case study of what can happen when development professionals build relationships that cross over social and political divides and challenge established power relations. They tell the story of how they worked together to lead a campaign to provide access to identity cards, and thus voting rights, to thousands of poor Bolivians in advance of a national election.[39]

Working from different social positions – one a development professional in a major bilateral aid agency, the other a development professional in a local NGO – Eyben and León developed a relationship that enabled them to challenge some of the ways of working of their respective organisations. As a result, they were able to roll out a campaign to provide access to identity cards in spite of vested interests that did not see the plan as a good use of aid resources.[40]

These case studies reveal that relationships are at the centre of development practice. At the 'interface' where different development actors come together, ideas for change are negotiated and co-created. Development professionals who are skilled at building relationships across boundaries can play a key role in opening new channels of communication and advocating for change.

Knowledge for development

Certain kinds of knowledge tend to be visible and valued in development planning; other kinds of knowledge are consistently overlooked. Local actors frequently have in-depth knowledge about what is required to improve their situation, or what works (or doesn't work) in particular local contexts. Yet this knowledge can be hard for outsiders to see or understand. Professionals value some forms of knowledge highly (such as objective scientific or policy knowledge). Yet they often devalue or fail to notice other forms of knowledge (such as subjective experiential or cultural knowledge).

Even among development professionals themselves, some kinds of 'development knowledge' are valued more highly than others – for instance, the knowledge of consultants may be seen as more valuable than the knowledge of staff, or quantitative studies seen to be 'better' than qualitative ones. Maia Green has observed that the knowledge practices in international development work strongly favour instrumental knowledge – that is, knowledge that can be immediately placed into practice. This makes it very difficult for development organisations to take on board knowledge that is open-ended and does not appear to have an immediate instrumental application.[41]

Over the years, anthropologists have created a large body of work on indigenous knowledge and argued for its practical importance to development work. Furthermore, many anthropologists have worked in a brokerage role to facilitate the coming-together of 'local' and 'outsider' knowledges to inform more effective development practice. The following case studies give some examples of the ways different kinds of knowledge are used – or ignored – in development work, and the practical impacts of knowledge inclusion and exclusion.

Case study thirteen: Local knowledge in projects

Development projects have traditionally been designed 'for' or 'on behalf of' beneficiaries by experts from elsewhere. Even when participatory processes are used, they often merely seek local people's *input* and *ideas* for projects, rather than acknowledge that local people have knowledge and expertise. Anthropologists, by contrast, have repeatedly called attention to the on-the-ground knowledge of people on the receiving end of development initiatives.

Anthropologists have demonstrated the relevance of 'indigenous' or 'local' knowledge on development topics as diverse as agriculture, health and governance. Nevertheless, the dynamic nature of local knowledge – which is often actively *practised* rather than simply possessed – may make these sources of knowledge difficult to recognise or take on board in projects. Further, stereotypes about the ignorance of people in rural or poor contexts

– and assumptions about the expertise of people from urban or wealthy contexts – can easily render local expertise invisible.

A case from Montserrat demonstrates how this can easily happen. In the Montserrat case, a development initiative aimed to rebuild two villages destroyed by natural disaster. Yet, in the end, the project failed to provide adequate housing for local people. When key decisions were made about construction of the new housing, the views of outside development workers prevailed; the developers, after all, were the experts. The in-depth knowledge of local people about their own environmental and social contexts was ignored.[42]

Local people knew from experience that a selected site for village rebuilding 'was too high for water pumping and too hard for properly sunk drains, but the development workers were the planners and surveyors'. Local construction workers knew that the prefabricated houses provided would not withstand hurricanes, and local women knew that their extended families would not fit into European-style houses. Local development actors had experiential, technical and cultural knowledge directly relevant to the rebuilding effort. Nevertheless, 'the houses were built without their input'.[43]

Development actors with power and social status – such as the planners and surveyors that came to Montserrat – are often assumed to have all the relevant development knowledge. Emma Crewe gives a compelling example: the case of development programs to design improved cooking stoves and distribute them to families in poor areas. Crewe details the range of actors involved in cooking stove projects: planners, engineers, artisans and, of course, the cooks. She then observed that the cooks themselves had little or no input into the design of the cooking stoves.[44]

Crewe asks a key question: *Why were cooks not leading stove development?* The answer was that the cooks were women, and they were poor. Their social positioning was such that other actors considered their views to be 'backward' and 'distorted by exotic traditional beliefs, in contrast to the apparently sophisticated, detached technical expertise of engineers and even social scientists'.[45]

These case studies provide an important caution for development professionals. When designing and implementing development projects, the knowledge of lower-status development actors can be key to success. Yet it is easily overlooked or ignored; while the knowledge of higher-status development actors has instant credibility – even when this is not entirely deserved. What men say, what professionals say or what foreigners from high-status places say can, in social settings, easily outweigh what women, non-professionals or people from low-status places say. Yet overlooking what lower-status actors know can lead to serious development mistakes.

Case study fourteen: Knowledge in context

Development projects, programs or policies frame development 'problems' and 'solutions' in certain ways. To define what needs to be done and how to do it, they draw on some kinds of knowledge, and exclude others. One common pattern is for developers to frame the problem or solution in a way that looks logical but overlooks other aspects of the local context that directly impact results.

A case study of a rural development project in the Pacific illustrates how a project can fail when developers lack knowledge of the local context. This promising project proposed to give landless peasants 20 acres of land, coconut seedlings and agricultural advice to start their own agricultural enterprises.[46] The project went on to deliver everything promised: including the square parcels of land. Unknown to the developers, however, local people reckoned their property boundaries differently: not in squares, but from the centre outward. Local systems of measurement may not have seemed relevant to a farming project, but when it came to implementing the project, no one understood the square plots. People could not work out where their 20 acres began and ended. In the end, the project failed due to persistent litigation over land boundaries.[47] This case demonstrates how cultural knowledge of the local context can be central to achieving practical development outcomes.

Incomplete knowledge about local context can have a range of practical consequences on the ground. For instance, an anti-erosion program in the Jbalan highlands of Morocco prohibited slash-and-burn agriculture and charcoal-making. For policy makers, prohibiting activities that encouraged tree-cutting seemed an eminently logical way to protect the environment: it would enable highland communities to retain trees and thus reduce erosion. However, far-away policy makers lacked key knowledge about the local environmental and economic context. As a result, their policy ultimately *increased* erosion – while significantly damaging the livelihoods of local people.[48]

First, removing charcoal-making as a livelihood option meant that many of the poorer residents in the Jbalan highlands were deprived of a key source of income. In response they were forced to leave to seek employment in the cities – not their preferred option, and one which further exacerbated rural–urban migration problems for the government. Next, farmers who were forced to abandon slash-and-burn agriculture had no alternative but to cultivate permanent fields on hill slopes instead. While the slash-and-burn process retained the roots of trees in the soil to prevent erosion while forests regenerated, hill slope cultivation was very prone to erosion – and gave much lower yields.[49]

Knowledge of the local context can also explain behaviours that may seem ill-considered or illogical to development workers. For instance, farmers may choose to plant lower-yielding traditional varieties of crops instead of high-yield modern varieties. This is not necessarily symptomatic of ignorance or stubborn clinging to the past. Often the traditional varieties are hardier, more resistant to the vagaries of local environments, have desirable characteristics (e.g. they taste or store better) and/or require fewer expensive inputs. A study of rice and maize producers in the Philippines, for instance, demonstrated how farmers' choice to combine traditional and modern crops and technologies made sense as a strategy to reduce risks, decrease turn-around time between crops and optimise scarce resources.[50] Looking at data such as crop yields in isolation from their context can miss these connections that are central to local livelihoods.

These case studies show how the knowledge that is relevant for development decision making often reaches far beyond the parameters of the development problem itself. In local contexts, different aspects of lives and livelihoods are interrelated in ways that are not necessarily visible to outsiders. Local logics can look quite different than project logics, but they can also reveal important sources of knowledge that are needed to enable successful change.

Case study fifteen: Knowledge brokering

The knowledge of poor and powerless people is regularly excluded from development decision making. Nevertheless, some development professionals recognise its value and try to find ways to bring it into dialogue with professionals and policy makers. Bringing very different knowledges and logics into dialogue across social divides is complex, however. It often requires skilful cultural brokering, as well as a place where people feel safe and encouraged to share what they know.

Case studies of participatory appraisal workshops in two countries – Uganda and Tanzania – by Johan Pottier provide an up-close look at knowledge exchange across social boundaries. In one case, participatory knowledge-sharing encounters served to break down social distance among participants and enable mutual learning. In the other case, an apparently similar process reinforced social distance and prevented people from sharing what they knew.[51]

In Pottier's Ugandan case study, skilled facilitators used strategic questioning and inclusive processes to broker an open dialogue among social

continued

actors from very different backgrounds. The facilitators recognised that different participants knew different things, and so provided opportunities for different kinds of knowledge to inform understanding of local farming issues. For instance, in one workshop, a male farmer shared his knowledge that early planting of millet could generate a more successful crop. A female farmer then explained that this strategy would only work for farmers with enough labour to manage the weeds that emerge with early crops. The wealthy male farmer's strategy was valid, but would not work for many of the people in the room, who could not access the labour required. An inclusive process that encouraged participants to speak, in a political context where they felt safe to do so, enabled different kinds of knowledge to come together across social boundaries.[52]

Conversely, in Pottier's Tanzanian case, the facilitators gave little attention to farmers' views, rather privileging the knowledge and analysis of government staff. While apparently very participatory, these workshops and field visits did not achieve much in the way of knowledge-sharing. The facilitators included farmers in the process but assumed they had little to say and discouraged them from expressing what they knew, deferring instead to the 'experts'. Nor did the political environment encourage villagers to speak up. As a result even apparently 'participatory' workshops only reinforced social divides and silenced knowledge exchange.[53]

These up-close views of participatory processes suggest that real knowledge-sharing is indeed possible across social divides, but it is highly dependent on the context in which the encounter takes place – and the skills of the facilitators or brokers that mediate the exchange. Brokerage across cultures is a challenging task; as Mette Olwig has shown in her study of development brokers in Northern Ghana, brokers do not merely translate: they must work across and within quite different – and sometimes contradictory – development logics. Practically, this means that a broker may simultaneously agree or disagree with the same statement, or present information that appears confusing or illogical.

Olwig gives the example of a local development worker who explained that a local group was both a 'women's group' and 'a group with men in it'. It was, in fact, both. According to external development logics, a 'women's group' was a group organised to provide women with access to resources. The group was therefore a women's group. According to local development logics, however, groups should include men so that men would support women to participate. The group therefore had several male members. The development worker was able to navigate across both professional and local knowledges to explain that the 'women's group' had 'men in it', but these knowledges did not sit comfortably together.[54]

For development professionals, recognising that different development actors know different things, and think about things differently, is a key first step to making development processes more inclusive. It is not enough for people to simply be invited to participate in a process; these case studies suggest that for mutual learning to occur, there must be opportunities for people to articulate what they know and be heard and respected by others. People who are comfortable working across cultures can often play an important brokerage role, helping very different social groups to understand one another across social divides.

Old institutions and new

Some institutions are powerful; they control resources and influence. These institutions are embodied in powerful organisations, such as national governments, international agencies, large corporations and international NGOs. People who understand these organisations' ways of working, and occupy influential positions in the relevant structures, have considerable room to manoeuvre. Those who do not may find themselves continually at a disadvantage.

Anthropological studies of planned development provide a number of practical insights about the role of institutions in development. These studies frequently observe the meeting points of more powerful and less powerful organisations and communities. They describe how the former's ways of working dominate these encounters, as well as the often savvy ways in which the latter respond. The meeting points of the institutions of the 'powerful' and the 'powerless' draw attention to the problematic nature of power relations in development, as well as opportunities for change.

One key insight from this work is that no development project or program ever starts on a blank institutional slate. In every social context where development professionals seek to work, there are already structures, rules and norms in force. Understanding 'how things work' now is the logical starting point for understanding what is actually required to create change.

Case study sixteen: McDevelopment institutions

Grassroots groups, small NGOs, and informal communities have their own ways of working. But when they work with powerful organisations, the latter's ways of working generally take precedence. Even when powerful

continued

development organisations aim to work in partnership with less powerful groups, their own ways of working nearly always set the terms of the development encounter.

Anthropologists Jeanne Simonelli and Duncan Earle studied the relationship between an Indigenous Mayan community in Chiapas, Mexico, a Mexican development NGO and its international funders. This ethnography demonstrated how the Mayan community's own local ways of working were systematically disregarded by outside developers. The developers, for their part, framed the local communities in accepted development categories – such as 'refugee women'. They then offered the community a selection of pre-packaged development project options that bore little relation to the community's own aspirations for change.[55]

The pre-set 'menu' of development project options offered by the NGO – which Simonelli and Earle termed 'McDevelopment' – sat within the dominant institutional apparatus of development assistance. It responded to the mission of the NGO and its funders and was structured to deliver popular development goals such as gender sensitivity and improved livelihoods. All the projects offered on the menu were small-scale production projects for women: vegetable gardens, hen raising and so forth. The women were invited to choose the most appropriate option for them in a reflective process – yet the process excluded any other options. The NGO's pre-set menu of women-only projects 'not only failed, but also divided the community along gender lines in the process'.[56]

Later, when the same women were given an opportunity to plan a development investment outside the dominant institutional frame, they came up with very different options, such as value-adding the local coffee crop, and investing in education. The institutions they used to arrive at these options worked quite differently than those of the developers. Decisions were made at community level through an extended process of conversation and consultation involving women and men, separately and together. The anthropologists observed how community institutions worked differently than dominant development institutions, using different processes and valuing different things.[57]

The case study is instructive. Development work is often framed by dominant institutional ways of working that are designed to deliver benefits in particular ways. Yet the people and organisations that developers seek to work with may work quite differently. Expecting people and organisations to change how they work to fit a standardised model overlooks their own – often functional – institutions and the options those institutions can provide for positive change.

Case study seventeen: Institutions and participation

Participatory development proposes that the beneficiaries of development initiatives become active participants, working alongside development professionals to define problems and implement solutions. Yet when development organisations seek to engage people and organisations in participatory and partnership-based approaches, this raises the questions: On whose 'institutional turf' does the development encounter occur, that of the developers or the participants?[58] Whose ways of working dominate the relationship? Anthropologists have observed that most participatory encounters still take place almost exclusively on the institutional turf of developers; developers create the spaces for participation and 'invite' others in.[59]

Anthropologist Elizabeth Harrison gives a compelling example. Her ethnography of older people's forums in Southern England provides detailed insight into a familiar process: community members are invited to participate in and contribute to the local authority's strategic planning processes. Yet the extent to which community participants were actually able to have a say in future planning was tightly pre-framed by the local authority. Many forum participants observed that in the end the invitation from the Council to have a say meant very little: 'because decisions have effectively already been made – somewhere else and by other people'.[60]

This is a disturbingly common pattern when participatory processes are observed up-close: they nearly always take place within the ways of working of established institutions, where there is little scope to create much change. Harrison described how the local authority's strategic planning process subtly reinforced social distance and power relations between decision makers and other residents. From the configuration of the room, to the arrival of dignitaries, to the re-writing of community priorities, participation was 'performed' as an invited process on the local authority's institutional turf. Local community participants were not passive puppets, however; they entered the process with their own agendas. Despite their limited ability to have much say, some learned how to engage more strategically with decision makers as a result of experiencing how the Council's institutions worked.[61]

Looking up-close at participatory process, as in this case study, reveals that institutional turf matters for participatory processes. The organisation that defines how the process will work also maintains considerable control over both the process and the outcomes of participation. It frames what questions will be asked (and what will be 'out of scope'), and defines the conditions in which communication occurs – including rules and norms about accepted behaviour and whose voices carry the most weight. Any effort to shift power relations in favour of marginalised groups must therefore consider the institutional turf on which encounters with these groups take place; in the end, whose ways of working dominate?

Case study eighteen: Toward inclusive institutions

In local contexts around the world, institutions already exist to provide many of the things that development professionals are interested in: education, health, production, infrastructure, market access and so forth. Equally, institutions can actively block people's access to these same development goods. Yet development organisations often have an incomplete understanding of the institutional context they seek to change: what the current institutions are, or where they have come from. This makes it difficult to grapple with the core problem of how to make institutions work better for disadvantaged groups.

David Lewis has argued that development projects and policies have a tendency to ignore the wider institutional context and history of the places where they work in favour of an imagined 'perpetual present'. He argues for the importance of taking a historical view of development contexts to gain a realistic picture of the nature of the institutions within which people work. For instance, he cites a project in Mozambique that completely ignored the history of the area – such as the recent civil war – when attempting to design rural development solutions. The technical solutions failed to work, because rural poverty was part of a political context where the ways of working limited people's access to resources.[62]

Current institutions directly affect what people can do – their room to manoeuvre. These institutions, in turn, often have long histories. Development efforts that fail to understand local institutional environments risk reinforcing the conditions that favour already-privileged groups, and disadvantage already-disadvantaged ones.

Efforts to make institutions more inclusive can easily backfire when current institutional environments are poorly understood. For instance, *gender mainstreaming* is an approach that aims to institutionalise the inclusion of women and their interests across-the-board in development work. It is a significant effort to make institutions more inclusive. Yet in practice, mainstreaming can serve to make women's disadvantage and exclusion invisible. By turning gender into a management category and ignoring how power relations are institutionalised in particular contexts, gender mainstreaming makes it 'easier to put real women and men, and the messy realities of their lives and relations, at a certain distance'.[63] Rather than make institutions more inclusive, a 'mainstreaming' approach may inadvertently reproduce exclusion.

At the same time, the institutions 'that people create for themselves' can hold the seeds of significant change.[64] Social movements that build on existing grassroots organisations can potentially challenge what Arjun

Appadurai calls the 'terms of engagement' between privileged and marginalised groups.[65] He gives the example of slum-dwellers' organisations in India that have mobilised at national level to engage with policy makers around the issue of affordable housing. These groups have created new spaces, such as housing exhibitions 'by and for the poor', which have allowed them to dialogue with policy makers on more equal terms, while showcasing their own knowledge and practices. Such community-driven institutional spaces are able to include policy makers in their processes, rather than waiting to be included by them.

For development professionals, these case studies emphasise the importance of paying attention to the institutions that are already present on the ground before launching into a change process. Understanding the history of a place can give important clues about current institutional arrangements and how they may limit people's room to manoeuvre. Further, paying attention to the institutions that people create for themselves can provide important insight about workable strategies for change.

Learning from development practice

The case studies in this section have shown that the people on the receiving end of development initiatives generally 'do' development differently than developers do. Understanding their perspectives and experiences can be the key to more effective change-making.

Anthropologists who study development pay attention to the people involved in development practice and what happens when they come together: including the relationships they build, or fail to build, in their quests for positive change. Regularly, the actors who design and implement development initiatives focus on problems and solutions rather than people and contexts. Often they overlook what their 'target groups' are already doing, and fail to recognise what they know about their local contexts and the constraints they face. Practically this means that development work often proceeds with incomplete or inaccurate knowledge – and this can have real impacts for people on the receiving end of development initiatives.

Even when development professionals take a participatory approach and work to build relationships and partnerships with those they want to help, their own professional frames can easily get in the way. It is startlingly common for developers to expect the people they work with to fit into their pre-set development categories, speak development language, feel comfortable in the institutional spaces of development organisations, and work in the same way that developers do. Professional frames can make some ideas and approaches seem 'as much a given as gravity', while obscuring other options that do not sit clearly in the frame. Locals, in turn, may feel awed by powerful outsiders, their expertise and the resources they represent. Locals may not feel encouraged or even safe to share what they know or how they already work.

Recognising that local people have expertise can help avoid costly errors and inappropriate 'solutions' to poorly understood issues. Recognising that local actors have their own institutional arrangements for getting things done can help ensure that change plans are realistic and sustainable. Further, bringing different kinds of knowledge and institutions into dialogue in development decision making can create opportunities for mutual learning – and even spark innovative ways to tackle development problems.[66]

Effective communication and relationship-building across social and cultural divides is not easy, but it is vital to effective development work. Powerful change-making can occur when development actors are able to recognise and question their own stereotypes, assumptions and comfortable framings of the world, and create new kinds of relationships across boundaries that challenge disadvantage and exclusion. A starting point for cross-boundary relationship-building is to recognise that less advantaged groups have their own knowledges and institutions. These represent different ways of seeing, thinking and acting; they often frame 'problems' and 'solutions' quite differently than developers do. Recognising and engaging with these knowledges and institutions can open up exciting pathways to change.

Summary: Lessons for practice

What can development professionals learn from reading the work of anthropologists? This chapter has illustrated a number of practical lessons that can be learned from anthropologists' up-close view of change processes – both planned and unplanned.

First, anthropology encourages us to be aware of development actors: diverse people, male and female, rich and poor, powerful and not, situated in diverse contexts, and actively involved in creating, navigating and coping with change.

As Olivier de Sardan has observed, development work:

> requires the involvement of numerous social actors, belonging to both 'target groups' and to development institutions. Their professional status, their norms of action and competence vary considerably. Development 'in the field' is the product of these multiple interactions, which no economic model in a laboratory can predict, but whose modalities anthropology can describe and attempt to interpret.[67]

Paying attention to development actors teaches us that change does not happen in a vacuum, but always in active engagement with people. Different people have different goals, different resources, and face different kinds of opportunities and constraints. People affect change processes, and are affected by them, differently. Development professionals ignore this at their peril.

Anthropologists show us that ill-considered development initiatives can too easily create problems rather than opportunities for already-disadvantaged actors. At the same time, they show us that people in all kinds of unexpected places are capable of driving change. Change is not the exclusive domain of development professionals. Paying attention to who 'grassroots actors' really are and what they do can reveal surprising allies for change.

Next, anthropology highlights the existence of 'local' or 'indigenous' knowledges, and the frequent tendency to overlook them. These situated knowledges have been developed out of long practical experience living and working in local contexts. Local forms of knowledge are important, but they are also very hard for outsiders to see – especially when the locals are stereotyped as ignorant or uneducated. Anthropologists show how people are often effective experimenters, innovators and managers of change over time. In the end, failure to recognise and take on board local expertise is at the root of many failed development efforts.

Because development initiatives always take place in particular contexts, local knowledge about these contexts always matters. Abstract expertise is never enough. Paying attention to indigenous knowledges pushes development work beyond its own frameworks. It teaches us that no one individual or organisation holds the answers to development challenges – nor can be expected to. Further, it teaches us that actors have different logics about change, and may take quite different approaches. When development professionals are willing to step outside their comfort zones and acknowledge local knowledges, they may find unexpected insights and ideas for change.

Finally, the case studies in this chapter have illustrated that institutions play a key role in shifting or reinforcing the relationships that reproduce disadvantage. Accepted 'ways of working' may reproduce power relations that place some groups at consistent disadvantage. Normal ways of 'doing development' can actually be part of the problem rather than the solution. When development professionals recognise how institutions currently work, they are better placed to shift them in more inclusive directions. In particular, the often-rich institutional resources of 'poor' and 'disadvantaged' groups themselves can create springboards for positive change.

This chapter has provided only a sampling of case studies from the rich anthropological literature on social and economic change. It has aimed to highlight some of the most obvious and practical lessons from this literature for improving the effectiveness of development work. The work of anthropologists provides a set of clear insights about why development programs and projects often fail to create positive change, and reveals untapped resources for approaching old development problems in new ways.

Anthropologists show us how to look beyond the surface of development projects and programs to see that people are always at the centre of the process. Anthropologists' approach to understanding change provides the raw materials to reframe how we 'do' development: from a technical process managed by experts, to a social and cultural process that is embedded in relationships. Interested readers are encouraged to explore further and learn more.

Notes

1 Gabriel (1991, pp.3, 33).
2 Other groups included the Canari federation in what is now Southern Ecuador; the Chibcha and Tairona states in what is now Colombia; and other local political domains and federations across the region (Wolf 1982, pp.59–65).
3 Ibid., pp.45–50.
4 Geertz (1963, pp.28, 82).
5 Hamilton (1998).
6 Leach (1992).
7 Warren et al. (1995).
8 Rusten and Gold (1995, p.89).
9 Thurston and Parker (1995, pp.140–143).
10 Ayers (1995).
11 Niamir (1995, pp.245–248).
12 Shaw (1992, p.214).
13 Long (2013, p.357).
14 Long (2013).
15 Fairhead (1993, p.193).
16 Richards (1993, p.62).
17 See e.g. Rhoades and Bebbington (1995) on experimentation by potato farmers in Peru, and Sharland (1995) on women's agricultural experimentation with a range of subsistence crops in Sudan.
18 Gamser and Appleton (1995, pp.343–344).
19 Ibid., pp.340–342.
20 Netting (1991, p.150).
21 Horowitz (1995, p.494).
22 Babb (1998).
23 See e.g. Eversole et al. (2005).
24 See e.g. Smith (2003).
25 Lee (2016).
26 Kahakalau (2016).
27 See e.g. Esman and Uphoff (1984).
28 Zorn (2004, p.162).
29 Attwood (1992).
30 Eyben (2006, 2014).
31 Horowitz (1995).
32 Fairhead and Leach (1997, p.37).
33 Peattie (1991, p.36).
34 Ibid., p.36.
35 Ibid., pp.38, 39.
36 See e.g. Standing et al. (2011).

37 Mosse (1997, p.278).
38 Mosse (1997, 2005).
39 Eyben and León (2006), Eyben (2014).
40 Ibid.
41 Green (2012, p.53).
42 Skinner (2003).
43 Ibid., p.104.
44 Crewe (1997).
45 Ibid., p.68.
46 Cochrane (1979).
47 Ibid.
48 Munson (1990).
49 Ibid.
50 Fujisaka (1995).
51 Pottier (1997).
52 Ibid., pp.208–210.
53 Ibid., pp.210–219.
54 Olwig (2013, pp.436–438).
55 Simonelli and Earle (2003).
56 Ibid., p.194.
57 Ibid.
58 Eversole (2012).
59 Cornwall (2008).
60 Harrison (2012, p.160).
61 Harrison (2012).
62 Lewis (2009); analysis of the Mozambique case study cited from Wrangham (2004).
63 Smyth (2010, p.148).
64 Quote from Cornwall (2008, p.275).
65 Appadurai (2004).
66 See e.g. Eversole (2015).
67 Olivier de Sardan (2005, p.28).

3

An anthropological framework for development practice

The first chapter in this book introduced a set of key ideas from the anthropology of development that have direct relevance for development work. The second chapter then presented case studies of what these ideas look like in practice, and why they matter in real places and communities around the world. Chapters 1 and 2 were about ideas and why they matter in practice.

Chapters 3 and 4 now look at how to put these ideas *into* practice. The next two chapters will show how ideas from the anthropology of development can be applied in development practice itself – in the everyday operational realities of designing and implementing development initiatives.

Anthropologists understand that designing and implementing development initiatives is a social and cultural process – a process with people at the centre. They understand that development work on the ground seldom plays out in the ways that those who plan it expect, and they are trained to see why. Anthropologists offer a very different way of seeing development work than economists, managers or policy experts do. Arguably, this people-centred framing of development is much better suited to grappling with contemporary development concerns like participation, diversity, equity and rights – issues which also have people at their centre.

Anthropology to date has had a surprisingly small impact on development policy and practice because anthropologists tend not to present what they know in a language or format that makes sense to busy development professionals. Development professionals are focused first and foremost on creating practical impacts, and they want user-friendly knowledge that they can use. Most development professionals do not have formal anthropological training; nor do they generally have time to read extensively. How can they apply ideas from anthropology in their day-to-day work?

The framework presented in this chapter is intended as a starting point. It distils key insights from the anthropological literature in Chapters 1 and 2 into a practical tool that development professionals, local leaders and others who are working for change can use. This framework enables people without anthropological training to embed an anthropological approach into their work. From there, Chapter 4 shows how this framework can be used in the context of day-to-day development practice, with reference to typical project and program stages, making use of familiar development tools.

This anthropological framework presented in this chapter is neither rigid nor prescriptive. Rather, it has been designed as a flexible tool that emphasises *reflexivity* or *reflexive practice* – defined as a process of actively paying attention to the ideas and ways of working that guide our own practice. The anthropologist Rosalind Eyben defines reflexivity as the process of deliberately 'being alert to different perspectives' and 'becoming unsettled about what is normal'.[1] It is the willingness to step outside of our own, comfortable cultural frames and try to view development from the perspectives of others.

This chapter names commonly accepted development ideas and practices which are part of the culture of how many development organisations work. It draws attention to these 'normal' ways of working and how they make it difficult or impossible to manage development as a social and cultural process. Through a reflexive process, this chapter seeks to creatively reimagine and reframe development practice with people at the centre, drawing upon the ideas already discussed in Chapters 1 and 2.

Framing and reframing development

Development work is about creating or catalysing change. Whatever the policy document, program proposal or project logframe, all development initiatives essentially boil down to this core aim: trying to create some kind of change.

Anthropologists argue that all change-creating involves people. Individuals, communities and organisations – in government departments, in community organisations, in multinational firms – are the forces behind economic and social change. They drive economic, social and environmental impacts.

Taking an anthropological approach to development practice means recognising that change is a social and cultural process. As a first step, this means putting people – rather than issues, problems, policies, projects, technologies or ideas – at the centre of development practice. An anthropological approach pays attention to how the interactions among people and their organisations influence the nature of change.

Putting people at the centre of development work is a significant shift. Most development professionals are trained to think very differently about creating change. We study issues, measure indicators and grapple with big ideas like poverty, economy, sustainability and governance. We identify promising technologies and ideas; design policies; implement initiatives; and report on the outcomes of our work. People are present, but almost as an afterthought. They are those *to whom* projects, policies, technologies or ideas are applied.

Putting people at the centre of development practice requires nothing short of *reframing* – changing the basic cultural frames that we use to think about development, and how we see our own roles in creating or catalysing change. Yet this reframing is urgently needed. Development professionals have been struggling for decades to take on board 'social' ideas about participation, gender, exclusion, Indigenous communities, ethnic minorities, stakeholders, partnerships and so forth. Despite considerable effort to incorporate these ideas into development practice, there is little real change. This is because *the dominant development framework is essentially incompatible with people-centred development practice*.

The dominant framework: Problems, targets, solutions

A *framework* is a useful practice tool because it shows what to focus on. No one can focus on everything at once. A framework highlights key ideas and categories; it is a way to make sense of complexity and hone in on what really matters for the task at hand. A simple representation of the dominant framework in professional development practice looks something like Figure 3.1.

In this dominant practice framework, a development professional identifies a development problem, a target group for intervention, and a desired solution of some sort. This is how he or she frames the problem. Once these three elements are clear, he or she crafts a strategy for achieving the solution. This strategy is based in an implicit or explicit theory of change; that is, a set of ideas about how change will happen.

When development professionals see development, they see problems, target groups, potential solutions and ways to achieve them. Like a metaphorical window-frame, the dominant practice framework (Figure 3.1) highlights the elements that are 'in view' when development professionals design and implement policies, programs or projects. These shared ways of seeing are echoed in shared ways of working.

The framework in Figure 3.1 presents a compellingly logical view of how development works: a problem or set of problems is defined. A target group is identified that experiences this problem. Perhaps they are farmers struggling with market access; families without access to fresh produce or clean water; or municipalities grappling with the challenges of good governance and efficient service delivery. Whatever the circumstances, development professionals are trained to define target groups and problems – not people and contexts.

This practice framework also has a pleasingly practical focus on solutions. These are usually technical solutions: for instance, tried-and-tested land-management technologies, cutting-edge ICT-enabled services, high-level organisational change strategies, or economic recipes for raising productivity.

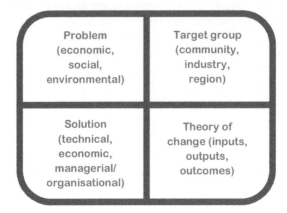

Problem (economic, social, environmental)	Target group (community, industry, region)
Solution (technical, economic, managerial/ organisational)	Theory of change (inputs, outputs, outcomes)

FIGURE 3.1 Dominant Practice Framework for Development Work

Development professionals create projects, programs and policies based on a technical prescription for what needs to happen in order to solve the identified problem. They mobilise expert knowledge via established development institutions to create change. Indeed, there seems to be little need for local people to know or do anything beyond what the experts advise.

The framework in Figure 3.1 summarises how many professionals see 'good' development practice: a process that is clearly structured, well targeted, led by experts and outcomes-focused. Inputs lead to outputs, which generate positive outcomes for the target group. If the outputs and outcomes are measurable, and an evaluation strategy is embedded into the framework to provide evidence of impact, this can even be called 'excellent' development practice. Indeed, with such a clear and logical framework guiding practice, it is surprising that the world should have any development problems left at all.

The view in the dominant practice framework described in Figure 3.1 is clear and compelling. This way of 'doing development' promises targeted action with a focus on results. Unfortunately, however, this view is also deeply inaccurate. Development initiatives seldom actually work this way in practice. Like a painted window, the view is attractive, but completely unrelated to what is happening on the other side of the glass.

The reality is that the entire process described in Figure 3.1 takes place on complex landscapes of people, organisations and institutions. The targets, problems and solutions that development professionals 'see' are just one small part of a larger landscape, and this larger landscape influences the outcomes in a myriad of ways.

For instance, the actors who comprise target groups usually have multiple identities and multiple concerns. Not all of them necessarily see that the 'problem' matters for them. When it does, it is often deeply intertwined with other problems and circumstances. The proposed solution may or may not be the most appropriate, and the proposed change process may be simply un-workable in the local context. At the same time, other actors and other agendas may intervene, taking change in another, unanticipated direction altogether.[2]

The view presented in Figure 3.1 is, uniquely, the developer's view. Only the developer, who is charged with creating or catalysing change, frames the development landscape in this way. Other actors have a different view of the development landscape: they see people, environments, relationships, connec-tions, contradictions. For developers, the centrality of development action – problem, target group, strategy and solution – is the frame that defines what they see, as well as what they fail to see.

It is no surprise, therefore, that development practice still struggles to integrate 'social' ideas like *participation* or *gender*. Too often, these ideas are simply tacked on while the business of development proceeds as usual. The target group may be encouraged to participate, but they are still positioned socially as the *target* group. Gender or ethnicity considerations can be integrated into the framework, but only insofar as they define the target group, problem, solution or preferred process. Stakeholders can be encouraged on board as partners, but only if they can work within the established frame.

Efforts to add in social considerations to the dominant ways of doing development generally fail to provide more than attractive window dressing. Even the best-meaning efforts to take social considerations on board fail to make any real change, because the social context is not visible within the mainstream development framework. The frame only shows 'social things' insofar as they fit into the development-centric worldview of target groups, issues, solutions and theories of change. Thus it becomes easy for development organisations to focus on 'women's development' separate from its community context, or to promote 'Indigenous development' using only non-Indigenous institutions.

An anthropological framework: Contexts, actors and resources

Putting people at the centre of development practice requires a significant reframing: it shifts the frame to focus on people, their organisations and the specific contexts in which they operate. Figure 3.2 suggests what an anthropo-logical framing of development work could look like.

Rather than pre-defined development 'problems', an anthropological frame-work focuses on the *contexts* where development occurs. Any context is a mix of interconnected problems and opportunities, enablers and limiters of change. In an anthropological framing of development, a change initiative is defined not by a *problem*, but by a context.

The context is what ultimately determines what the problem (or opportunity) actually is, and how it relates to other problems and opportunities on the landscape. Focusing on context thus avoids dividing development action into

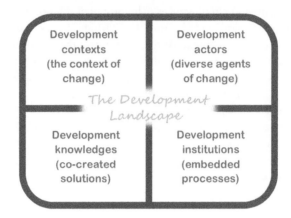

FIGURE 3.2 An Anthropological Framework for Development Work

sector-based silos that miss key connections – for instance between environments and livelihoods. It also guards against the tendency to overlook contextual factors that directly influence outcomes on the ground – such as the nature of local institutions.

Next, an anthropological framework focuses on *development actors* rather than 'target groups'. Target groups are always pre-defined according to the particular interests of the developers. It is therefore easy for developers' own stereotypes and assumptions to guide the categories in which they choose to place other people. An anthropological approach, by contrast, considers how the people and organisations on the landscape define themselves.

An anthropological framing also moves from passive to active voice: people are not passive 'targets'; rather, they are seen to be actors with agency. The erstwhile 'targets' of development are recognised as diverse actors, differently situated with reference to social characteristics such as gender, ethnicity, ability, rank, economic status or age. Further, these actors are aware that change initiatives may improve or worsen their situation and the situations of other people they care about. They act accordingly.

Finally, the anthropological framework in Figure 3.2 focuses on the *knowledges* and *institutions* of diverse development actors, seeing these as central to any change process. Anthropologists reveal the existence of a range of knowledges beyond the expert knowledge of professionals, and institutions beyond the dominant institutions of development practice. Rather than pre-packaged solutions and abstract theories of change, the emphasis in this framework shifts to co-created solutions and embedded processes of change.

Rather than focusing narrowly on expert-led solutions, an anthropological framework calls attention to the multiple, indigenous and local knowledges that can be mobilised to co-create more innovative solutions. And, rather than seeing change as a technical process that is managed by professionals in development institutions, an anthropological framework focuses on how local ways of working can create change that is sustainable over the long term. In the quest for development solutions and strategies, these diverse knowledges and institutions are key resources.

The framework in Figure 3.2 represents a practical way to reframe how we think about development work. It shifts the focus from *problems* to *contexts*; from *target groups* to *development actors*; and from technical solutions to a creative engagement with diverse knowledges and institutions. Rather than seeing development initiatives as operating in isolation from their context, an anthropological approach understands that all planned change takes place on a larger 'development landscape'. This landscape will, inevitably, influence how change happens and who benefits. The following sections describe the key components of the framework in more detail.

The development landscape

Figure 3.2 illustrates what an anthropological approach to development practice could look like. Some development professionals and organisations are already very comfortable with this kind of contextualised, people-centred approach to development work. They design projects, programs and policies with attention to people and context. Nevertheless, the dominant framework in development practice is still very much as described in Figure 3.1. Mainstream development policies, programs and projects revolve around problems, target groups and expert-driven theories of change that promise to deliver predictable solutions regardless of context.

The key difference between Figures 3.1 and 3.2 is that Figure 3.1 is opaque – it focuses only on the key elements of a development intervention, ignoring everything that sits outside the development 'frame'. Figure 3.2, by contrast, is transparent. It recognises that development initiatives always take place in particular social and physical contexts that will invariably influence the outcomes. Figure 3.2 thus sees development interventions in terms of their relationship to the broader *development landscape*.

The development landscape can be defined as the *overall setting in which any development action occurs*. It includes specific contexts, diverse actors, what they know, and the institutions they use to get things done. It is, therefore, a useful shorthand for many of the ideas discussed in the first two chapters of this book. On the development landscape, actors with different knowledges and ways of working interact to create or resist change.[3]

Development professionals working within the mainstream practice frame pay little attention to this landscape. As a result, they generally fail to anticipate the range of things that can happen (and often do happen) to sidetrack, hijack or reinvent the best-laid development plans. They are trained to look at the mechanics of the intervention specifically (as per Figure 3.1), and not to see it in relation to the people and situations on the broader development landscape (as per Figure 3.2).

The metaphor of *landscape* emphasises that development does not take place in a vacuum; nor do interactions among people and organisations ever remain within the boundaries of any particular development initiative. Rather, development actors act and interact in multiple ways, in and beyond specific initiatives. Their actions occur in observable physical places – from the forest floor to the board room – and from within social relationships that influence their room to manoeuvre and how they seek to influence change.

Attention to the development landscape allows us to start to unpack the 'black box' of policy making and question the elegantly planned lines of program design. It challenges the myth that good project or program management, or good policy implementation, is all that is required to make change happen. It allows us to see what really happens in development work: from the bird's-eye view of an international development agency, to the on-the-ground perspective of a local community leader. What do people and organisations actually do and why – and what do these actions and understandings reveal about the possibilities for change?

Attention to the development landscape enables change-making processes to start from what is actually there – the existing production system, the current political arrangements – rather than a developer's assumption of what is, isn't or should be there. It thus provides a realistic starting point for change. Further, it forewarns development professionals against the dangers of stumbling blindly onto a poorly understood development landscape, where well-meaning actions can very easily create unintended effects. As a tool for reflexive practice, an anthropological framework encourages us to always see our own work with reference to people and contexts and to ask: *Where do our initiatives fit on the development landscape?*

Development in context

Development policy decisions are often made from afar. This 'big-picture' view of problems and solutions tends to miss much of the nuance on the ground. As a result, poor fit between development initiatives and on-the-ground realities is common. In the early twenty-first century, there is a clear awareness that top-down decision making has limited effectiveness in achieving development outcomes. A key lesson from past failures is that the local context really does matter for effective development work.

Every development context is different: it has a different physical landscape, different social characteristics, and different institutions and cultural frames are in force. Of course there are commonalities across contexts, and some generalisations that can be made. But while these generalisations may usually hold true, they do not hold true everywhere. Just because smallholder farms *tend* to be managed in certain ways does not mean that they are *always* managed like that. Just because people *usually* seek certain goals does not mean that they *always* do. Different situations and considerations can widen or narrow people's choices and alter the room to manoeuvre that they have.

Realistically, it is possible to invest years in really understanding a particular development context. Yet development professionals seldom have the luxury of much time. They may be able to manage to visit local communities, but they cannot realistically understand everything about the contexts where they work. Nevertheless, being unable to know everything is not an excuse for knowing nothing. A surprising amount of development work is initiated without the decision makers ever having so much as visited the places or communities they seek to change. Ignoring context is perilous – it is a proven recipe for mistakes and unexpected consequences.

Generally, developers start with a problem that concerns them (as per the top left corner of Figure 3.1). Because development problems tend to repeat themselves in lots of different places around the world, there is a certain amount that developers already know about any given problem. They know about the problem – but they may know very little about what it looks like in different places. Focusing on a *problem* immediately narrows the field of vision. Other problems, which may be interrelated – and other assets and opportunities – are obscured from view.

Reframing the focus of practice from *problem* to *context* (as per the top left corner of Figure 3.2) avoids the basic problem of divorcing problem-solving from context. Rather than starting a change process with an abstract idea (*hunger, disadvantage, unemployment*) and what we know about this problem in a general sense, it starts instead from a concrete physical and social reality where the problem may – or may not – manifest. In doing so, it seeks to understand how the problem that is the topic of concern relates to other problems, assets and opportunities.

Development work is commonly structured around problems to be solved, but it is not difficult to reframe it. Three questions can be used in reflexive practice to guide the transition from *problem* to *context*:

1   *Is this a significant problem here, in this context?*

2   *If so, how does it connect to other problems and opportunities?*

3   *What enables and/or inhibits change in this context?*

(See Figure 3.3.)

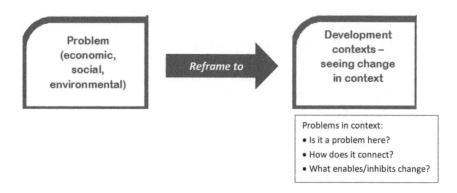

FIGURE 3.3 Reframing Problems in Context

Asking these questions does three important things:

1   The first question avoids the danger of assuming that there is a problem where none exists. This is surprisingly common. A serious issue in one context is often a complete non-issue in another, even in the same country, industry or type of community. The environment, economic base or social institutions may be quite different. Equally, what developers construe as a serious problem may not be all that serious, particularly when compared with other problems people are facing.

2   The second question acknowledges that development problems never occur in a vacuum, but are interconnected with other aspects of physical and social settings. The failure of children to attend school may be related to economic pressures, health issues, transport access, social disenfranchisement or something else. The symptom may look the same, yet the root problem may be quite different. Asking 'how' a problem connects to other problems and opportunities recognises that a problem can have different root causes, and thus, different potential solutions. It allows development work to address causes rather than merely symptoms.

3   The third question recognises the dynamic connections between a problem and other parts of people's lives. Asking 'What enables and/or inhibits change?' reveals opportunities for integrated development solutions in unexpected places. How is women's health linked to land tenure and local governance? What is the relationship between the incarceration of minority youth and the structure of justice institutions? Why might transport infrastructure be central to understanding unemployment? Finally, how are people themselves trying to create change in their own contexts, and how much room to manoeuvre do they have to do so?

The problems that we regularly address in development work never exist as things-in-themselves. As the case studies in Chapter 2 illustrated, development problems are always connected into the broader contexts in which people live: the nature of the physical landscape, the institutional options that are in force, the social roles that people occupy, the historical conditions of disadvantage that reproduce disadvantage in the present. Problems are located on physical and social landscapes – geographies of plenty and want, structures of privilege and disadvantage – which vary from place to place and from community to community.

Further, *problems* are intertwined with assets and opportunities. While starting with *problems* takes a deficit view, starting with *contexts* reveals the things that are working well. It makes assets visible – such as functioning local institutions, strong community leaders or in-depth local knowledge – that can provide a springboard for positive change. Seeing assets rather than problems places the focus on what is already working well and creates a positive starting point for action.

Taking *contexts* instead of *problems* as the starting point for development work is a simple yet significant reframing. It allows us to avoid the dangers of trying to solve 'problems' that we only partially understand, and shows us the real-world connections that point to where the deeper problems really lie. Attention to context forewarns us where off-the-shelf solutions are unlikely to work, and reveals assets and opportunities that are always present on the ground, but often just out of view. Reframing to see them can spark solutions in unexpected places.

Unpacking actors

Development actors are diverse, savvy and situated in particular contexts. While mainstream development work starts with pre-defined *target groups* (as per the top right corner of Figure 3.1), an anthropological framework shifts the focus to *development actors* (the top right corner of Figure 3.2). This anthropological framing recognises that a range of actors are involved in creating, resisting and negotiating change.

Focusing on *development actors* resolves one of the deep tensions that plague development work: the struggle to try to incorporate questions of participation, stakeholder engagement, social diversity and so forth within the mainstream development framework. The mainstream development framework is centred around developers' view of problems and solutions; in this framework, only developers truly act. Other actors, even when they 'participate', are still situated on the receiving end, positioned as 'targets' rather than actors in their own right. From within this framework, authentic stakeholder engagement and participation are essentially impossible to achieve.

An anthropological framework recognises that development professionals and policy makers are not the only actors on the landscape. Other actors also work to create change. Seeing a landscape full of actors with agency, rather than developers and passive target groups, reframes the development process itself. 'Whom will we work with?' is not presumed; indeed the question may be usefully flipped to 'Who will work with us?'

Further, actors are diverse, while target groups look homogeneous. It is easy to name a particular group as the 'target' of a development initiative, but the very act of naming it obscures the diversity within it. People in any given target group differ – by gender, ethnicity, nationality, religion, profession, education, place of residence, politics, life experiences (refugees, the unemployed, parents) – and in many other ways. Their varied identities and social positions mean that members of the same 'target group' may have markedly different interests, resources and room to manoeuvre.

Moving from *target groups* to *actors* is more than a question of language: it represents a significantly different way of seeing the people and organisations who are involved in a development process. It upends the assumed social positions: developees as passive recipients, developers as active helpers. These social positions have become so entrenched that some actors have even begun to see themselves as target groups, passively receiving external help. Changing the focus from *target groups* to *actors* allows us to see that these are indeed actors with agency – but being a savvy actor may mean 'playing along' with the categories of developers, if that is the only way to access the resources that they control.

A reflexive practitioner can easily shift the focus of development work from *target groups* to *actors* by asking:

1   *Who are the relevant people and organisations in the context where we are working?*

2   *What are they doing already?*

3   *How are they positioned socially; what will change mean for them?*

(See Figure 3.4.)

Focusing on actors rather than target groups does several important things:

1   The first question recognises that development processes and initiatives always take place in social settings, where there are usually many more relevant people and organisations than developers anticipate. They will have different perspectives on any proposed change process, as well as a range of pre-existing relationships with one another. Developers frequently seek to partner and engage with stakeholders to drive change, but they seldom scratch the surface of the day-to-day social relationships that reinforce business-as-usual.

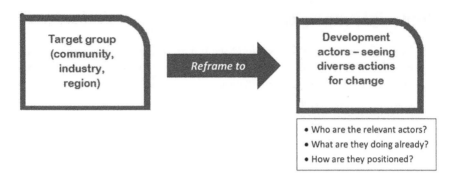

FIGURE 3.4 From Target Groups to Actors

2   The second question recognises that multiple people, communities and organisations have agency, or the ability to create change. Rather than putting the developer at the centre of the frame, responsible for 'doing it all', an anthropological framework considers the range of actors on the development landscape who may be potential allies (or opponents) in a change process – and who may already be working toward a solution.

3   The third question recognises that all actors have a particular social positioning that affects the kinds of resources that they can access, and the influence they can mobilise. These social positions – being a woman, or gay, or university-educated, or from a certain ethnic group or family – may mean a great deal about what you can or can't do, what resources you can access and how others treat you. Some actors are more powerful than others because their social positions give them better access to political, social, physical and financial resources and influence. Asking how actors are situated with reference to a change process can help to anticipate who is likely to end up better off – or worse off – as a result.

Starting with actors rather than pre-defined target groups places the focus of development work on real people, their situations, and how they affect and are affected by development processes. The case studies in Chapter 2 demonstrated that local actors and their organisations already drive change; yet they also showed how easy it is for these actors to be overlooked in expert-led development processes. Attention to actors reminds development professionals that they do not work in a social vacuum, and they do not need to create change on their own. There are allies (as well as opponents) in unexpected places, and untapped opportunities to mobilise relationships for change.

Knowledges and logics

Different development actors don't see things the same way. The way that an extension professional views a problem, and the way an experienced farmer views it, can be quite different. This is because they come at the problem from different cultural perspectives, knowing different things. The 'logical' response for each is not necessarily anything alike. Each may have important insights, but about different things. Like the old metaphor of the blind men and the elephant, different development actors perceive the problem – and its solutions – from different angles.

Despite extensive documentation of indigenous and local knowledges and their relevance to development work, development decision making is still dominated by a relatively narrow set of professional knowledge-sets. These occupy a privileged position; professional knowledge in fields like engineering, agronomy and economics is respected and trusted to provide development solutions. For professionals, taking on board the non-professional knowledges of development actors from different social and cultural backgrounds can be one of the most challenging aspects of an anthropological approach. At the same time, it is potentially the most rewarding.

Development knowledges can be defined as the different kinds of knowledge that are needed to address development issues. Anthropological research has shown that professional knowledge is often inadequate on its own to understand development problems. Professional knowledge tends to be highly rigorous but also highly focused; it is often ill-attuned to the particularities of local contexts, and poorly connected to other realms of professional or local expertise. When development initiatives have failed on the ground, it is often the concrete, contextualised insights provided by local and indigenous knowledge that have been missing.

Rather than relying on professional knowledge to provide technical solutions to development challenges, an anthropological approach draws upon multiple development knowledges. It moves away from narrow, expert-led technical solutions (the lower left corner of Figure 3.1) to a focus on multiple development knowledges and co-created solutions (as per Figure 3.2). This approach incorporates situated local knowledges alongside rigorous professional knowledges to understand and address development issues from different angles.

The following questions can be used in reflexive development practice to shift the focus from expert-led solutions to more inclusive knowledge processes:

1   *Who knows something about this?*

2   *What do they know?*

3   *How does that change how we see the problem and/or solution?*

(See Figure 3.5.)

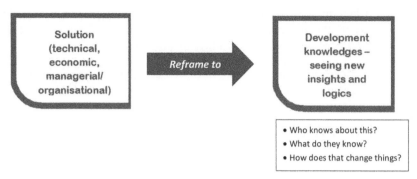

FIGURE 3.5 From Technical Solutions to Co-Innovation

Reframing to focus on multiple development knowledges significantly increases the knowledge resources that development practitioners can draw on in their quest for practical solutions. Specifically:

1   The first question recognises that many different development actors can contribute directly and productively to development solutions. Knowledge for development exists in unexpected places: across sectors, across localities, across social divides. In particular, anthropologists have shown that less advantaged groups have their own knowledges and logics which are not always shared by the professionals who aim to help them. What people know about their own realities is important, and recognising this knowledge is a vital first step in ensuring that development solutions are based on a real understanding of issues and needs.

2   The second question recognises the insights that can be gained from engaging with different development actors. Asking 'What do they know?' goes well beyond simply collecting information, to seek to understand the logics that inform people's choices and their beliefs about what is possible. Problems look different when seen from different perspectives. The logics underpinning our work are not necessarily shared by everyone we work with; and a logical strategy in one context is not necessarily logical in another. For instance, managing risk may be a more logical solution than increasing production in contexts where there are few economic or social safety nets. Asking what others know can help to avoid significant development mistakes.

3   Finally, the third question recognises the power of knowledge to transform practice. In particular, asking 'How does this change things?' shifts the focus away from expert-led solutions and creates space for solutions to be directly co-designed with the people and organisations who have traditionally been on the receiving end of outside expertise. An anthropological framing challenges the power relations that are embedded in development work

when professional knowledges dominate and local voices are silent. Respectful dialogue among different forms of knowledge is not only more inclusive – the solutions are likely to work better.

Development professionals are interested in solutions – this is, after all, the practical aim of what we do. A willingness to acknowledge, hear and take on board different kinds of knowledge paves the way to new kinds of solutions. Bringing different kinds of knowledge into dialogue makes it possible to co-create solutions that are technically sound and also make sense in local contexts.[4] Consider how some of the stories in Chapter 2 might have been different if local and professional knowledges had come into dialogue. What sort of new solutions might have emerged in forest management, rural development or post-disaster housing?

This process of opening up dialogue among different kinds of knowledge is not easy; it can quickly take professionals outside of their comfort zone. Asking what other people know, and considering how this changes things, can require questioning stereotypes and assumptions that we did not even know we had.[5] This process arrives at the sharp end of reflexive practice: a willingness to listen, question, and continually reassess the categories and logics we use to guide our work. Yet, by doing so, we can help to co-create solutions that are not only technically workable, but which actually work well for those we aim to help.

Institutions and change

Institutions are structures, rules and norms that guide how things are done. Established institutional arrangements benefit some people and disadvantage others – and they tend to change slowly. History matters deeply for development work, because historical relationships continue to echo in institutions today. Why do certain countries have seats on the UN Security Council and others do not? Why are certain families represented in political leadership generation after generation, while others have never cast a vote? If the underpinning relational logics of institutions are not questioned, old inequities can persist well into the present.

Development work aims to create practical, positive change. The mainstream framework for development work (as per Figure 3.1) proposes that change can be planned and delivered by mobilising certain inputs and activities for certain target groups, to create certain outputs and, ultimately, desired outcomes or solutions. Mainstream development approaches thus see change as a largely technical process that can be planned and managed to create targeted results.

The problem with this approach is that it only sees the particular behaviour or indicator that it is seeking to alter. This view is incomplete. No matter how compelling the theory of change or the evidence suggesting that doing *this* will lead to *that*, change processes never occur in a vacuum. All change takes place in institutional contexts. These institutions, grounded in history, will influence what actually happens. Institutions shape what people can do and how they can do it.

Whether or not a change takes place does not just depend on the inputs provided and the theory of change, but also on the extent to which the institutional context enables or inhibits this change. Take, for instance, the idea of using microfinance and enterprise development as a strategy to empower women. A woman may be offered inputs like finance and training and be interested in the opportunity to run a business; yet if the institutional environment makes it difficult for a woman in her social position to attend classes, accept credit or be publicly accepted as a business owner, then these inputs – regardless of their quality – will not be of much help in creating the desired change.

An anthropological framing recognises that institutions guide the choices and room to manoeuvre that people in particular social positions have. Institutions provide structures, guidelines and rules (more or less rigid) around how and when things are done, and by whom. Trying to change a behaviour (like farming practice) or an indicator (like maternal health) without understanding the institutional contexts that farmers or mothers currently work within will tend at best to generate a short-term change that does not survive the life of the development intervention. At worst, these initiatives may put the beneficiaries in a risky position as they try to push the boundaries of accepted practice to meet multiple expectations.

Seeing change as a technical process in which inputs lead neatly to outputs and outcomes (as per the lower right corner of Figure 3.1) overlooks the ways that institutions enable or inhibit change. Reframing change with attention to institutions (as per Figure 3.2) means recognising that change is a social and cultural process, not merely a technical one. People and organisations already have established ways of working that they use to get things done. New, changed ways of working must be at least marginally compatible with old, established ways of working, or they are unlikely to be taken on board. For instance, traditional healers or locally based extension staff are more likely to find ways to embed new practices locally than professionals or organisations that suddenly arrive on the scene aiming to change things. Building upon current institutional practice creates the groundwork for sustainable change.

Development practitioners can reframe their approach to change from a technical process to a social and cultural process, by asking the following questions:

1   *How do people and organisations currently work? What institutions are present?*

2   *How might they be constraining change, especially for historically disadvantaged groups?*

3   *How might they enable sustainable change?*

(See Figure 3.6.)

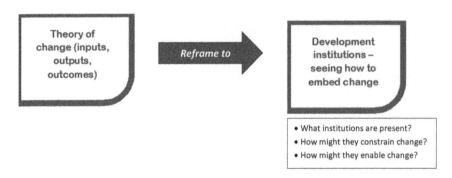

FIGURE 3.6 From Managing Change to Sustaining Change

Paying attention to the role of institutions shifts the focus from 'managing' change in a vacuum, to enabling change processes that have the potential to be self-sustaining over time:

1   The first question asks what institutions are already present in a particular context for doing the sorts of things developers are interested in: delivering education, for instance, or organising production. Current ways of working, and their historical reasons-for-being, are often invisible to outside developers. Yet these institutions are the starting point for future change. Rather than assuming that development work takes place on a 'blank slate', the first question starts the change process with an appreciative understanding of what is already there. In doing so, it forces consideration of how desired change 'fits' with the way things are currently done.

2   The second question recognises that current institutions may limit the room to manoeuvre that different development actors have. Structures, rules and norms can place significant constraints on what can be done, by whom and how – and they often have deep historical roots. This question seeks to identify the institutional constraints on actors, so as not to place them in untenable positions. It also draws attention to the ways that current institutions may serve to actively reproduce disadvantage for certain groups of people – so that development efforts do not unwittingly reinforce the same ways of working.

3   The final question recognises that current institutions can also enable change. One of the mistakes most commonly made by developers is to assume that local institutions do not exist or that they are necessarily inferior to those that arrive from elsewhere. They then import completely new ways of working that make sense to them, but are unfamiliar and often incompatible in the local context. Because of this, many development initiatives fold up as soon as the developers leave town. Starting with the

institutions that are already present creates the potential for more sustainable change that is deeply embedded in established ways of working. To the extent that change can be pursued incrementally and in familiar institutional ways, it is more likely to make sense to people, and to become part of 'business as usual' over the long term.

An anthropological reframing of development work recognises that institutions are central to any change process: change never happens in a vacuum. Institutions are already present on the ground in every development context. The case studies in Chapter 2 showed that even very disadvantaged social groups have their own institutions that they use to get things done. Yet these are often completely overlooked or even undermined by other development actors, who prefer for interactions to occur on their own institutional turf. Paying attention to institutions can reveal why development efforts that aim to help disadvantaged groups often simply reinforce current patterns of disadvantage. At the same time, noticing the institutions that are already present in local contexts may suggest new ways of working, and new ways to embed change.

Summary: Doing development anthropologically

This chapter has shown how the ideas from anthropology presented in Chapters 1 and 2 can be used in a practical way to inform development work. These insights challenge the mainstream way of seeing and doing development as illustrated in Figure 3.1. They enable us to reframe development, as per Figure 3.2, as a social and cultural process, revealing how planned development efforts take place upon a larger development landscape.

Development landscapes are complex, and anthropologists are skilled at describing complexity. Yet complexity is hard to manage in practice. The framework presented in this chapter is intended to distil complex insights from the anthropology of development into a simple tool that can be used to inform day-to-day development practice. This is a tool for recognising complexity and making it manageable within the practical demands of development work, above all by encouraging reflexivity and a willingness to see things differently.

Figure 3.2 reframes development practice with reference to a development landscape that includes:

1   the *contexts* in which change takes place;
2   the various *actors* who drive development action – people and organisations that are diverse, savvy and situated, with differing amounts of power;
3   the *knowledges* and *logics* of different actors – recognising the importance of bringing these knowledges and logics into dialogue; and

4   the *institutions* already present on the landscape, and how they enable or constrain change.

An anthropological approach refocuses the mainstream view of development, which is centred on the actions, logics and institutions of developers. It replaces this with an anthropological understanding of how social and economic change really works.

Rather than conceiving of change as a technical process in which the problems of target groups can be defined and solved in isolation from the rest of their lives, an anthropological approach recognises that problems need to be understood *in context*. Understanding how and why local context matters in problem-solving can help to avoid costly mistakes in the design and implementation of development initiatives. Further, attention to context can reveal assets, opportunities and positive starting points for change.

An anthropological approach also shifts the focus of development work from *target groups* to *development actors*: recognising that actors drive change. While the definition of 'target groups' may simply reflect outsiders' stereotypes and assumptions, attention to actors considers the diverse social positionings from which people and organisations seek to influence change. An anthropological approach to development work seeks to understand what actions different people and organisations are already taking, and the different ways that change is likely to affect them.

An anthropological approach also recognises that multiple *knowledges* and *institutions* are already present on the development landscape. Different ways of seeing, knowing and doing can be difficult to recognise within development practice, where developers' ways of seeing, knowing and doing tend to dominate. Nevertheless, these knowledges and institutions can provide important resources for change. Developers who are willing to be reflexive about their own practice can begin to recognise and appreciate the knowledges and institutions of others, and to explore how these can reframe 'problems' and 'solutions' in new ways – opening up space for innovative co-created solutions.

In place of an opaque picture of carefully managed change, an anthropological framework provides a clear view onto real-life development landscapes where development professionals work alongside other actors, who are themselves agents for change. Anthropological work challenges the illusion that a well-designed program or policy will run to plan regardless of real-world situations. It questions the assumption that simply tacking on 'social' considerations to mainstream ways of working will result in more inclusive results. Rather, an anthropological approach puts people at the centre of development work, in all of their diversity. It encourages development professionals to work within the development landscape, rather than in isolation from it.

The anthropological framework presented in this chapter is intended to encourage reflexive, people-centred development practice: practice that takes on board lessons learned from past failures, while seeking out opportunities for future innovation. At the heart of the framework is reflexivity: being willing to listen and learn from others, respect their ways of working, and recognise that relationships for change extend well beyond the limits of any particular development initiative. This approach to practice not only helps to avoid costly errors; it can reveal unexpected knowledge, allies and resources. The framework in this chapter suggests a simple way for development professionals to approach our change-making role anthropologically, to appreciate the development landscape on which we work, and to navigate it confidently.

Notes

1   Eyben (2014, p.20).
2   The anthropologist Jean-Pierre Olivier de Sardan (2005) refers to this as 'sidetracking'.
3   Olivier de Sardan's term 'development arena' is a similar concept: the *arena* is where development actors interact. He notes that, 'A development enterprise is always an arena in which various logics and strategies come into confrontation: those of the initiators of the development enterprise confront those of the so-called target population' (2005, p.137). While useful, the metaphor of an 'arena' focuses only on project-specific interactions, while 'landscape' emphasises their interconnections with broader social processes and physical places.
4   Eversole (2015) describes this process in development practice as *knowledge partnering*. This same principle has also percolated into other fields, for instance in the form of user-centred design.
5   Chambers (1983) gives some great practical examples of stereotypes and assumptions that he regularly observed in rural development work in the 1970s and 1980s. Many, unfortunately, persist today.

4

Applying the framework: Tools and approaches

Ideas from the anthropology of development can be applied in practice to support more effective, inclusive and innovative development work. Chapter 3 presented a framework for development work using an anthropological approach. It drew together anthropologists' findings about the importance of contexts, actors, knowledges and institutions in development processes, and showed how these ideas can be applied directly in development practice.

This chapter continues to explore how development professionals can integrate an anthropological approach into their day-to-day development practice. It introduces the reader to a set of practical tools that can be used to apply the elements of the framework in their work, at each stage of a typical project or program cycle: from assessment and design through to implementation and evaluation.

These tools will not look unfamiliar to development professionals. Most of the tools in this chapter are commonly used and are not particularly anthropological in themselves. They can, however, be used anthropologically, to provide a deeper understanding of contexts, actors, knowledges and institutions in development initiatives. These tools can be used even when working within significant time and resource constraints, and in organisations that are not particularly attuned to anthropological approaches.

Practically, it is important to recognise that most development initiatives continue to be designed and implemented within a mainstream practice framework. They focus on problems, target groups and technical solutions: because this is the way development is usually done. Further, there are often significant resource and time constraints in day-to-day development practice. Yet professionals who work in these contexts can still choose to adopt an anthropological approach. With only a little effort, they can begin to shift the focus from problems to contexts; from target groups to actors; and from externally driven to co-constructed change processes and solutions.

This is an incremental approach to change. It recognises that most of us must work within organisations whose own ways of working are not particularly anthropological, and which are unlikely to change fast. Development institutions tend to be self-sustaining systems; there is strong pressure to maintain the status quo. Pushing institutions beyond their comfort zones is often impossible; they simply do not 'take on board' ideas that challenge their established ways of working.[1] Many development organisations are structured around the idea that success can be designed and delivered for others. Nevertheless, even very small adjustments can create marked improvements in the ultimate outcomes of their work.

This chapter is structured around some of the most common institutional features of development projects and programs. Essentially, there are certain things that development organisations of all kinds nearly always do. When starting a process of change, they first *assess* the current situation. Then they *design* some kind of initiative or intervention which aims to improve the situation. From there, they *implement* the initiative or intervention.

Traditionally, assessment, design and implementation were all that was required. However, this left unanswered the question of whether or not the initiative actually succeeded in doing what it set out to do. When time and money have been invested, it is important to know whether – and to what extent – the initiative actually worked. As a result of growing concern with development effectiveness, more organisations now *evaluate* their initiatives. They report their outcomes, and ideally, they mobilise the lessons learned from evaluation to inform the design of future initiatives, creating a basic action–learning cycle.

Assessment, design, implementation and *evaluation* are core processes in development organisations of all kinds: from local government councils creating a municipal economic development framework, to multilateral development organisations rolling out food security programs across several countries. All organisations have their own priorities, funding regimes, governance systems and operational structures. Yet, despite these obvious differences, all assess, design, implement and – to some extent at least – evaluate their change-making work.

In the context of these common development processes, there are a number of tools and approaches that development professionals can use to integrate an anthropological approach into their work. These tools can be used within typical time and resource constraints, and without formal training in anthropological methods. The following sections describe a cross-section of tools for assessment, design, implementation and evaluation, and how they can be used to approach development work more anthropologically, with people at the centre.

Assessment tools: Understanding contexts

In the real world, change is an ongoing process. In development work, however, change has a starting point. Change starts when someone identifies a problem or issue that needs to be addressed, or a situation that needs to be improved. Thus, most development efforts start with some sort of *assessment* of the current situation or issue. Assessment creates a 'baseline', a starting point for future change.[2]

In practice, there are many ways to conduct an assessment, and many tools are available to assist the process. This section will provide an introduction to some tools that can be used to conduct anthropologically informed assessments: that is, assessment that pays close attention to the social contexts of change.

Perhaps one of the most common comments by anthropologists involved in practical development work is that anthropological knowledge is often brought into projects and programs too late, after initiatives have already been designed.[3] Bringing anthropological knowledge in to inform development work from the start can help to avoid serious errors and pitfalls. Taking an anthropological approach to assessment can reveal key aspects of the local context – history, environment, logistics, power relations, root causes of surface problems – and use these to orient a more effective and responsive project or program design.

Assessment in general can be said to vary along three dimensions. Assessments can range from:

1   the cursory to the detailed;

2   the focused to the integrated; and

3   the elite to the participatory.

It is disturbingly common for development initiatives to be based upon only *cursory* assessments, with little or no evidence about what is really going on in a particular context. Often, organisations jump from 'problem' to 'solution' – often, a clever-sounding solution borrowed from elsewhere – with only a sketchy understanding of what the problem actually is. Projects or programs may be implemented simply because someone in power thinks they are a good idea and is willing to fund them. Often, simple anecdotal observations (*too many young people on the streets*) paired with stereotypes and assumptions (*idle young people cause problems, jobless people lack skills*) convert into a compelling assumed 'need' for development interventions where no real need exists.

By contrast, other development initiatives start with *detailed* assessments of the current situation. These assessments take the time to understand what is really going on, before any development initiative is designed. Assessments may collect extensive baseline data, and undertake in-depth needs assessments with organisations and communities. In some cases, professional anthropologists are hired to conduct assessments. Detailed assessments take time, but they ensure that the project team has a nuanced understanding of the contexts in which change will take place. Most assessment exercises sit somewhere in between very cursory and very detailed – located somewhere along the spectrum that balances detail and evidence with time and cost.

A second dimension of assessment is the extent to which the assessment takes a *focused* or *integrated* approach. At one extreme of the spectrum, a focused assessment exercise looks narrowly at a particular sectoral issue: for instance, maternal health assessments or industry skills assessments. Highly focused assessments can provide targeted information for those working in a sector. At the same time, they are framed to look only at specific, pre-defined indicators, rather than the things that influence health or skills. At the other end of the spectrum, highly integrated assessments look across sectors and issues to spot connections and causal factors, and to identify assumptions influencing how issues are being defined. Place-based or community-based assessments often take an integrated approach to understanding issues in context.

A third dimension of assessment is the type of knowledge they draw upon. Assessments range from those informed exclusively by *elite* knowledge, to those which are highly *participatory*. On the *elite* end of the spectrum, assessment is conducted by topic experts or consultants drawing on a limited range of accepted data sources. The perspectives of the chosen experts frame the content of the assessment. On the *participatory* end of the spectrum, the assessment process draws upon multiple kinds of knowledge, beyond commonly accepted data sources. A range of people and organisations from different backgrounds are involved in the assessment, including those from less powerful groups, and their knowledge is brought into dialogue across boundaries.

These three dimensions of assessment – detail, focus and inclusiveness – may vary enormously from one assessment exercise to another. There are a range of approaches to assessment, sitting at different places along each spectrum. Thus, there is considerable scope to take an anthropological approach.

An anthropologically informed assessment will invest time in gaining a clear understanding of the context from the start. Learning as much as possible about the context at the beginning of any process 'reduces the risk that false assumptions creep into the design of development programmes'.[4] Further, an anthropological approach will aim to give particular space to understanding the viewpoints of marginalised groups and those who are the 'targets'. An anthropological approach to assessment will thus tend toward the detailed (rather than cursory), integrated (rather than tightly focused) and participatory (rather than elite) end of each spectrum.

Nevertheless, compromise is required in practice. Detail, integration and authentic participation take time; and in resource-constrained development organisations, time is often in short supply. In practice situations, the prevailing resource pressures are nearly always toward the cursory, focused and elite ends of the assessment spectrums: rapid, targeted assessments by expert consultants based on readily available data. Detailed, integrated and participatory assessments to gain an in-depth understanding of particular contexts may be seen as desirable but too costly to be feasible.

Development practitioners who want to take an anthropologically informed approach can navigate this tension by demonstrating that even some contextual information is much better than none at all. Further, contextual information need not be costly to obtain. Even when there is 'no time' for extended fieldwork or in-depth face-to-face consultations, a range of assessment tools can be used to shed light on particular contexts, tap into local knowledge, question elite assumptions, and reveal obstacles and resources that would otherwise remain invisible.

Time spent on good assessment is frequently time saved in avoiding problems, mistakes, wasted resources and duplication of efforts down the track. The following sections give some examples of assessment tools that can be used to support practical, anthropologically informed assessments, even in resource-constrained environments.

Tools for desktop assessment

Due to time constraints, as well as the rapid availability of data in the information age, it is not uncommon for many assessments to take place at desk level only. Such assessments often rely heavily on official statistics and documents prepared by sector-based experts. Few resources are, however, available at desk level that capture local knowledge or the perspectives of marginalised groups. Nevertheless, an anthropological approach to assessment is possible, even within the constraints of desktop work.

Stakeholder mapping. Stakeholder mapping is a common approach to identifying key actors and their relationship to an organisation or development process. Stakeholder mapping can be used in development work at the assessment stage to identify actors and their likely 'stake' or interest in any planned change process. Most commonly, stakeholder maps visualise how external actors relate to a given organisation, and identify which actors are more or less important to achieving organisational aims. An anthropological stakeholder map moves the development organisation out of the centre of the picture. Instead, it puts the focus on the relationships among the various actors in a given context, where the development organisation is one actor among many. This provides a structured way to identify the organisations and groups that are already working in a given context and how they are positioned socially and politically, with attention to power relationships and alignment or conflict among them. Mapping stakeholders from the desktop has clear limitations, though: it is generally not possible to identify all relevant actors and their relationships from afar. Nevertheless, stakeholder mapping provides a way to move beyond catch-all categories like 'industry' and 'community' and start to make a concrete assessment of who in a given context may contribute to or challenge change.[5]

Knowledge scan. Desktop assessments are typically based on easily available published data and reports. This creates a strong bias toward codified knowledge from established sources. These sources may be largely self-referential and so can perpetuate various forms of 'development ignorance'. Knowledge scan is an anthropologically informed tool for identifying a broader range of knowledge sources about a given topic. A simple grid (see Figure 4.1) can be used to identify both codified and non-codified forms of knowledge about a topic in a particular context. The knowledge scan focuses on actors – from academic experts to grassroots organisations – and the different kinds of knowledge each can bring to inform an understanding of the topic in context. Knowledge scan thus takes an intentionally inclusive view of the knowledge sources that are needed to inform an accurate assessment.

Desktop consultation. Even when working from the limits of the desktop, assessment exercises can seek input and insights from a range of actors. Telephone, video calls and email can be used to communicate with and learn from people and organisations that have been identified via stakeholder mapping

Knowledge Scan		
Topic, Context, or Issue	**Who Knows about it – Formal Knowledge** _From datasets, academic publications, policy documents, practice literature, websites, etc._	**Who Knows about it – Informal Knowledge** _From sector experts, fieldworkers, local organisations, community leaders, service users, etc._

FIGURE 4.1 Knowledge Scan Grid

or knowledge scan. As always, the desktop has clear limitations: not everyone has access to communication technologies, and some conversations can only really happen face to face, requiring time and trust. It is important to be honest about what – and who – is being missed out in desktop consultation. Still, an email relationship is better than no relationship, and a video chat generally better than no communication at all. Any available technology can be used to start conversations and pose questions, with the emphasis on respectful relationships and learning as much as possible from the people and organisations who know their context well.

Tools for assessment workshops

When assessment moves from the desktop to the field, it is still often time-constrained; brief site visits and workshops are common. Site visits can give a sense of a place and its people; they are valuable and yet limited. As critiqued decades ago by Robert Chambers, carefully staged site visits may constitute little more than development tourism.[6] Workshops are a useful approach because they bring together a lot of actors at once; they are thus time-effective and, at least on the surface, have the potential to be participatory. There are, however, many ways to conduct a workshop, and many agendas that can underpin it: from simply rubber-stamping decisions already made, to providing a rich environment for dialogue across boundaries. The following tools can be used with skilled facilitation to provide opportunities for actors to dialogue with each other and with outside professionals in assessment workshops.

Appreciative enquiry. Appreciative enquiry focuses the assessment first and foremost on what is already working well. It thus 'flips' the common tendency for development conversations to focus on problems or things that need to change. Instead, appreciative enquiry starts with the positive. The advantage of appreciative enquiry is that each context is understood on its own terms, recognising its uniqueness, strengths and the assets that are already present. Problems and issues are acknowledged, but so are resources and assets. Appreciative enquiry therefore avoids the common mistake at assessment stage of overlooking the key knowledge-sets, organisations and institutional resources that are already present in local communities. Appreciative enquiry workshops can also incorporate other strengths-based assessment tools such as asset mapping.

SWOT and Gap analyses. It may seem odd to include tools originally developed for business analysis in a toolkit for anthropologically informed development work. Yet tools like SWOT analysis and gap analysis are practical assessment tools that start from the assumption that people are knowledgeable about their own context and what they want to achieve. They are thus compatible with an anthropological focus on actors and their knowledges. SWOT is simply an acronym for 'strengths, weaknesses, opportunities and threats'. It provides a structured way for participants in an assessment workshop to articulate their own assessments of their current situation. A simple SWOT grid can be used to encourage discussion of both current strengths and weaknesses, and what these mean in terms of opportunities and threats going forward. While the SWOT focuses on 'where we are', gap analysis describes the space or gap between 'where we are' and 'where we want to be'. Gap analysis thus starts to move beyond assessment to suggest a potential agenda for change. SWOT and gap analyses can be used to tap deeply into local knowledge and understanding of current situations and desired change trajectories, so long as workshop participants share broadly similar situations and common interests.

System mapping. There are numerous variations of system mapping tools, from the mapping of value chains for particular products, to the mapping of power brokers and how they influence local political systems. What all of these different approaches to system mapping have in common is that they seek to reveal linkages and connections: they are thus situated at the 'integrated' end of the assessment spectrum. System mapping tools can be used in assessment workshops to open up in-depth dialogue and analysis of how 'this' relates to 'that' – identifying the key influencers and contributors to particular situations or outcomes. The effectiveness of such in-depth exploration of relationships will vary depending on the quality of facilitation, the political sensitivity of the topic, and the nature of who is in the room (and what they are willing to share). Still, efforts at systems mapping provide a valuable way to move beyond a surface assessment of contexts to explore drivers and root causes of change.[7]

Community-based assessment tools

The first two groups of assessment tools generally assume that an external development professional or group is leading the assessment process. Nevertheless, some assessment exercises are committed to the participatory end of the spectrum. They take community-based approaches to assessment which are largely directed by local people themselves. Of course, a commitment to being participatory and community-based does not necessarily preclude the process being co-opted by external developers or by local elites. Community-based approaches may or may not achieve participation for everyone. Still, the tools included below can be used to encourage a range of people to become involved in assessment activities, and to communicate their in-depth knowledge of their local contexts to each other and to outsiders.

Participatory Rural Appraisal (PRA). Participatory rural appraisal (PRA) is a basic and probably the best-known variant of community-based assessment. While originally developed as a way for professionals to make an on-the-ground assessment of local conditions by undertaking a low-cost 'Rapid Rural Appraisal' (RRA), approaches to appraisal have evolved over the decades to have a stronger emphasis on local people conducting their own assessments. PRA encourages local people to articulate their own knowledge about their local context. It employs a number of hands-on tools that people can use to analyse and communicate what they know to others. Some of these include stakeholder and system-mapping activities that can be undertaken by people with limited literacy, and communicated across language divides. While often still focused on communicating local knowledge to external decision makers, PRA may also take an action–learning approach, using local assessments as a starting point for reflection and the design of community-driven change processes.[8]

Participatory visual methods. Community-based assessment tools often encourage a process of reflection on the current situation as a precursor to exploring opportunities for change. Participatory visual methods can be a particularly powerful way of encouraging both individuals and groups to reflect on their current situation, by moving beyond numbers and words to 'show' what things are like in a particular context. Popular visual methods include participatory video, photo elicitation or photovoice (using photography to express experiences or aspirations) and digital storytelling. These methods can be used in a range of ways, but they generally involve individuals and/or groups recording their observations or experiences in a visual form and then reflecting upon these and their significance, sometimes in the form of a narrative or story. Assessment processes using visual methods may be externally facilitated or self-organising; and they may or may not seek to communicate the results to external audiences.[9]

Participatory statistics. Development policy makers have a noted preference for 'hard data' to inform decision making. Participatory statistics is a way to provide hard data from a community base, in a way that is context-sensitive and relevant. Participatory statistics mobilise people who are directly familiar with their own local contexts to collect quantitative data on indicators that matter to them. They thus offer a potentially powerful community-based assessment tool. Recent technologies such as smartphones, dedicated apps and cloud technology have opened up new options for data collection and data sharing, so that the users of infrastructure, resources or services can immediately record data based on their observations and upload them into a shared dataset. This can create a way for a large number of people to participate in assessing the quality, condition or availability of current infrastructure, resources or services.

Design tools: Crafting actions for change

Development work is the intentional pursuit of change; thus, the design of initiatives to achieve change is a central feature of all development work. This section presents some tools that can be used to encourage anthropologically informed design of projects, programs and policies.

Design, like assessment, can vary according to different dimensions. The design of development initiatives can range from:

1   closed-ended to open-ended;

2   narrow to broad;

3   single- to multi-actor.

First, the design of a development initiative can be closed-ended, open-ended or somewhere in between. Closed-ended design is tightly scoped around inputs, activities, outputs and outcomes. Initiatives at this end of the spectrum have clear, pre-defined measures of what 'success' will look like. At the other end of the spectrum, initiatives with an open-ended design are less prescriptive about what will happen during the change process. Open-ended design provides a more flexible framework for action that can be refined and altered over time, in response to changing conditions or the different priorities of actors.

Next, the design of development initiatives varies according to whether the focus of the proposed change effort is narrow or broad. Narrow design focuses on a single issue and a targeted solution. Projects and programs can become very targeted indeed: for instance, 'raising school attendance for ethnic minority girls' or 'providing a social procurement toolbox to municipal managers'. Moving along the spectrum, other development initiatives are designed with attention to the connections among issues, such as educational access, attainment and economic opportunity. At the broad end of the spectrum, 'integrated' project and program design seeks to move beyond issues to understand the dynamics of contexts: for instance, in the case of integrated rural development.

Finally, the design of development initiatives can be located anywhere along a spectrum from top-down design by a single actor, to very participatory, multi-actor design. At the single-actor end of the spectrum, project design is undertaken by a single decision maker or organisation, typically 'for' or 'on behalf of' others. This type of project design is limited to a single view of change. Moving along the spectrum, a larger number of development actors becomes involved in the project design process, bringing a range of perspectives. In the most participatory design approaches, many different knowledge-sets inform the project design, including those of intended beneficiaries.

An anthropological approach favours design approaches that are at the open-ended, broad and multi-actor ends of the spectrum. Open-ended design provides the most space for diverse actors to navigate complex real-world contexts and negotiate different visions of change, while a broad approach is more able to take into account contextual factors and relationships among different aspects of people's lives. Multi-actor design is more likely to incorporate the perspectives of beneficiaries and what they know about their own situations.

Nevertheless, as with assessment, there are compromises in practice. The more open-ended the design of an initiative, the more flexible and responsive it will be; but it can also be more difficult to get funders on board or ensure accountability. The broader the design, the more the initiative can be contextualised in real-world systems; yet broad initiatives are more complex to manage, and there is greater risk of trying to achieve 'everything and nothing'. Finally, the more actors who are involved in design, the more inclusive and potentially innovative it will be; but the process will be correspondingly complex to manage, and conflicting imperatives may emerge that are difficult or impossible to resolve.

An anthropological approach to design can navigate these tensions by recognising that while some compromises may be necessary, it is preferable to deal with complexity and conflict at the design stage, rather than coming up against immovable and costly obstacles when implementation is already underway. Development initiatives – projects, programs and policy – never take place in a vacuum; they 'hit the ground' on development landscapes populated by other actors with their own ideas, concerns and initiatives. An anthropological approach recognises this, and encourages development professionals to 'design in' these actors and their actions, rather than wait for problems and conflicts to occur down the track.

Numerous tools are available from the community development and participatory development traditions that can be used to bring different actors together to design and co-design initiatives. A small selection is provided below.

Visioning and planning tools

Visioning and planning tools help people and groups to articulate and share their own development visions and logics with each other, with other groups and with outside professionals. These tools can be used to open up creative conversations about what desired change looks like and what is needed in order to achieve it.

Community visioning workshops. Community visioning workshops bring together people with common interests to articulate their desired futures. Visioning workshops aim to articulate a common desired 'vision' for the future that can be used as a basis for collective action. Various tools and techniques can be used to capture the visions of individual actors, but the aim is typically to generate consensus: for instance, by using ranking and voting techniques to determine the top priorities for the group as a whole. While potentially a very grassroots approach to designing development initiatives, visioning workshops need to be carefully facilitated so that the voices and perspectives of less powerful or vocal actors in the group are heard and taken on board. As always, social positions and power dynamics will influence who comes, who speaks and what they say. Participatory PowerPoint can also be used in workshop contexts, in place of old-school butchers' paper or whiteboards, to keep a running record of key points on a screen that everyone can see and alter as required.

Participatory conversations. While community visioning workshops often start by asking participants to visualise their desired future state, participatory conversations provide a more open-ended way of initiating a planning process. Participatory conversation workshops have no pre-set agenda – for future change or otherwise. Rather, they aim to place the entire agenda in the hands of participants, who generate their own list of topics to discuss. Two of the best-known participatory conversation methods are the SOSOTEC approach (self-organising systems on the edge of chaos) developed by Robert Chambers, in which participants use cards to brainstorm agenda topics; and World Café and other forms of 'Open Space Technology', where participants self-organise into table groups around topics of common interest.[10] Participatory conversations are only lightly facilitated, as they essentially aim to provide a space in which people can self-organise to discuss and act upon the topics that are important to them.

Scenario planning. A key aspect of designing development initiatives is the need to plan for what is essentially an unknowable future. Some changes may be under the control of those designing an initiative, such as the decision about where to invest funds or build a piece of infrastructure; but other changes will be outside their control altogether, such as the behaviour of weather, markets or other development actors. Designing development initiatives therefore always involves a significant amount of grappling with the unknown. Scenario planning is a tool for recognising and managing the fact that there are change dynamics that are at play which cannot be known, but which may nevertheless be anticipated to some extent. Scenario planning tools are a structured way to imagine different future options in light of both their likelihood – given current trends – and their desirability for different actors.

Participatory approaches to scenario planning are particularly useful in that they provide opportunities for a range of people – including those with in-depth local knowledge – to contribute to the design and assessment of future scenarios. These scenarios can then provide the basis for considered action. When the design of initiatives involves real, physical places and landscapes, visual tools such as participatory geographical information systems (GIS) can be used to represent different land-use, infrastructure or urban-design scenarios. By enabling a range of people to suggest different options and then see what these would look like in a particular context, participatory GIS can facilitate knowledge-sharing and effective co-design.

Participatory Impact Pathways Analysis (PIPA). PIPA is another workshop-based tool that brings together a cross-section of actors to discuss how a project or program can be designed to create change. PIPA is useful at the design, implementation and evaluation phases of a development initiative, as the PIPA process encourages stakeholders to articulate 'impact pathways' which can then be embedded in the monitoring and evaluation (M&E) design and tracked throughout the project or program. The PIPA approach recognises that different development actors or stakeholders have different theories about how change works. Workshop processes encourage them to articulate problems and their causes, a project vision, a network map of actors, and to produce an 'Outcomes Logic Model' that specifies what each group of actors needs to do in order for change to occur.[11]

Resourcing tools

Visions and plans are important; but actually moving from ideas to action requires resources. Money, people and other resources are generally required to launch a change initiative. Thus, resourcing is always a key concern at design stage. Yet, few tools are available to help.

Traditionally, resources for development have been sourced from governments, from international cooperation and from NGOs and philanthropic sources. A *written proposal* has typically been the accepted tool for attracting resources, whether through an internal organisational or government process, an external grant round or a direct invitation to submit. Proposals take a variety of forms, but typically they propose a problem, solution, approach, timeline and budget. The development proposal is thus a key tool that is regularly employed at design stage to attract resources. By its nature, however, the proposal tends to perpetuate a view of development centred on problems, target groups, and the illusion of manageable, time-bound development solutions.

Contemporary development work is, however, changing. Even as traditional development resources are increasingly constrained, new resourcing arrangements are appearing on the landscape. There are now a wide range of financial resourcing arrangements for development work, from crowd-funding by the public, to direct community-to-community funding (for instance via migrants' hometown associations), social investment schemes, and a range of other multi-stakeholder and place-based partnerships that combine resources across sectors. Further, even traditional development resourcing arrangements are changing, as new innovations like participatory budgeting – which involve citizens directly in public-sector budget decisions – challenge old ways of allocating resources.[12]

An anthropological approach recognises that resources always come from actors, and all actors have interests and agendas. The interests and agendas of those providing and receiving the resources may or may not match up. For instance, public–private partnerships can bring important financial and management resources to public development efforts; at the same time, private-sector actors have their own agendas. Thus, it cannot be assumed that public–private arrangements will necessarily benefit citizens who cannot pay. Corporate or government partners may bring incredible knowledge, networks and financial assets to community organisations, but they can also dominate the relationship and turn it to their own ends. Partnerships are always comprised of actors, whose interests may or may not be aligned. At the design stage, it is important to consider the nature of the relationships among resourcing partners.[13]

Further, an anthropological approach to resourcing draws attention to the range of actors who may be able to contribute resources to a change initiative. Many actors who could potentially provide resources simply fly 'under the radar' and are overlooked. Resourcing does not always come from the expected places or take monetary form. Development actors from a range of backgrounds may contribute other important resources such as time, expertise, infrastructure, materials, information, and political or market connections. Involving a range of actors and their resources at design stage can mean that development initiatives are better-resourced in the long run.

The following questions can be used to orient a resourcing strategy at design stage that recognises actors and resources, and configures them creatively – each refers back to some of the techniques already outlined in this chapter:

- Who are the stakeholders here? Who wants change (and so might invest time, energy, stuff or money to help us achieve it)? (*Refer to Stakeholder mapping.*)

- Who – people, organisations – would know about what different kinds of resources are available and how to access them? (*Refer to Knowledge scan.*)

- What might different stakeholders be willing to give, do, recommend or influence to make this idea or initiative happen? (*Refer to Desktop consultation.*)

- What is already present on the landscape that is working well, and that we could build upon – rather than necessarily starting from scratch? (*Refer to Appreciative enquiry and SWOT analysis.*)

- Where could this idea or initiative be providing a 'missing piece' or solving a problem for someone else? (*Refer to Gap analysis and System mapping.*)

- What resources and knowledge are already present in local communities and don't need to be replicated? (*Refer to Community-based assessment tools.*)

These questions are not exhaustive, but they begin to reveal the presence of resources – and allies – in often unexpected places.

Doing development work in resource-constrained environments is not necessarily the severe disadvantage it is often portrayed to be. While wealthy organisations can afford to perpetuate waste and duplication, resource-constrained organisations must be proactive and find resources where they can. Often, this involves finding ways to engage and work together with others across boundaries – moving beyond personal and organisational comfort zones, challenging old stereotypes, silos and assumptions, and taking on board new knowledges and insights. Resourcing can therefore be a key touch point for innovation in development work.

Implementation tools: Life on the development landscape

Anthropologists generally observe a wide gulf between planning and implementation in development work. What is expected to happen when projects are designed, and what actually happens, may bear little resemblance to each other. Development initiatives are not implemented in a vacuum; they play out on complex social landscapes. What actually happens is determined in the interactions among development actors with different agendas, logics and institutional ways of working.

Unlike assessment and design, it is therefore hard to describe implementation on any kind of spectrum. Implementation is dynamic and often highly unpredictable. While there are institutional norms in development work that encourage development efforts to look well planned and predictable, a great deal of what actually happens is improvised and opportunistic, regardless of how it is presented in final reports.

An anthropological approach to implementation pays attention to the day-to-day interactions among actors – individuals, groups and organisations – and how these interactions determine what actually happens when projects, programs and policies hit the ground. The dynamics of relationships, at least as much as the structure of the initiatives, influence the results. Over the years, experienced development practitioners have evolved tools to manage relationship dynamics in development work. These tools can be used to incorporate an anthropological approach to implementation.

Participatory governance tools

Participatory governance tools provide mechanisms for project, program or policy implementation to be responsive, over time, to the needs of diverse interest groups. Recognising that implementation is a dynamic and often unpredictable process, these tools provide a way to get a wide range of actors involved in decision making, adaptive learning and managing uncertainty.

The effectiveness of participatory governance arrangements varies; in some cases these arrangements are dominated by elites, or set up as symbolic but powerless bodies with no real say over decision making. Yet, at their best, participatory governance arrangements can be a way for a range of different actors to actively guide the terms of the development encounter, bring different knowledge-sets to bear on decision making, and ensure that initiatives meet the needs of diverse groups.

Steering groups. Steering groups are perhaps one of the most common governance tools used in the implementation of development initiatives. 'Steering groups' is the generic term; they may also be called steering committees, consultative groups, advisory bodies or any of several similar names, and they may vary in size from three or four people to a dozen or more. All such groups have the basic function of providing advice and direction on the implementation of a project, program or policy. They generally comprise a cross-section of individuals from different backgrounds and areas of expertise, to ensure a mix of skills and perspectives. Often, steering group members are chosen with the aim to 'represent' the interests of particular organisations, communities or social sectors. An anthropological approach to steering groups recognises that individuals vary in the extent to which they are actually able or willing to represent the interests of others – particularly when they are supposedly 'representing' very diverse communities or sectors. Nevertheless, despite this limitation, steering groups are a valuable mechanism for bringing together a cross-section of actors with different knowledges and networks to guide the implementation of initiatives over time.

Local development partnerships. Local development partnerships bring together a range of organisational partners to implement an initiative. The focus is typically on initiatives for a defined geographic area, such as a neighbourhood, town, city or local region. The number and type of partners involved vary; ideally, such partnerships will cut across social divides to bring together both elite organisations such as municipal governments and chambers of commerce with grassroots and informal community clubs and groups. The idea of a local development partnership is, quite simply, for organisations to work together across sectors to create change in the local area. In practice, however, local development partnerships vary considerably in terms of the amount of actual participation from partners, especially less powerful partners. Collective Impact is one approach to local development partnerships that has attracted interest in practice circles.[14] The essential premise of Collective Impact is that organisations can have a larger impact on social issues by working together. Nevertheless, in practice, Collective Impact partnerships are often limited to different service-providing organisations, with limited or no participation from service users.

Devolution and auspicing. While steering groups and local development partnerships aim to bring together development actors from different sectors to guide the implementation of initiatives, other tools focus on empowering less powerful groups to take charge of implementation directly, using their own organisations and ways of working. In devolution, decision-making power and resources are devolved to local organisations, which are then responsible for implementing initiatives at the coalface. For instance, an initiative may be devolved to a local council rather than being centrally managed from a national government office; or contracted out to a locally based organisation rather than relying on fly-in project managers from far away.

Auspicing is another arrangement that can be used to enable less powerful groups to directly implement their own initiatives. Often these groups are informal and do not have the legal credentials to access grant funding, loans or other resources such as rental premises. An auspicing relationship with a more established organisation can be used to provide a way for such groups to access resources that would otherwise remain out of reach. Auspicing generally involves providing a legal guarantee for the group; auspicing organisations may also provide other forms of support such as access to specialist advice, mentoring or office infrastructure.

Participatory management tools

A number of tools have been developed over the years to manage the implementation of development initiatives. Many of these management tools, such as logframes or program logics, are premised on the idea that initiatives can be implemented in a straightforward and predictable way. An anthropological approach urges caution when using these tools, because of the way that actors and social dynamics ultimately shape the outcomes of initiatives. The following tools provide ways to position people and relationships more centrally in the implementation process.

Decentralised management. One strategy that can be used to deal with the social complexity of implementation is to take a decentralised (or devolved) management approach. Decentralised management shifts the responsibility for management decision making from high-level managers in central offices to the people who work at the 'coalface' in local contexts. Those staff who are most familiar with the local conditions in their own area are therefore empowered to take responsive decisions. The role of high-level management becomes one of coordination and strategic leadership rather than direct management of the implementation process in every context. Resourcing and reporting tools provide guidance, but also flexibility for local staff: for instance, by focusing on evidence of outcomes rather than prescribing specific activities or outputs. A decentralised management approach draws on the local knowledge of local staff, enabling large, multi-site organisations to become more agile and responsive to diverse local conditions.

Embedded engagement. While development organisations often seek to engage with marginalised groups during the process of project or program implementation, an alternative is to be engaged by them. Contemporary Web 2.0 technologies make it relatively easy and cost-effective for organisations to embed engagement into their day-to-day activities, by providing easy ways for people to contact them and provide input and feedback on an ongoing basis. Tools that make development organisations and governments easier for outsiders to engage can provide an important communication channel between powerful organisations and those on the receiving end of initiatives. Ideas and suggestions can be 'fed in', and complaints and problems aired in real time, providing opportunities for continuous improvement. Of course, the usefulness of these kinds of embedded engagement approaches ultimately depends on the extent to which the technology is accessible for the people concerned, and whether powerful organisations are able or willing to 'hear' and take on board the ideas and concerns of their constituencies.

Participatory monitoring. Participatory monitoring tools can be used to enable partners and intended beneficiaries to track the progress and/or outcomes of an initiative. Data is collected on the ground as initiatives roll out, and then shared among stakeholders to provide detailed and up-to-date feedback about the initiative and what it is achieving (or failing to achieve). Mobile monitoring, for instance, can be used to track variables such as water health or infrastructure conditions as they change over time. Community score cards (discussed later in this chapter) can be used to provide data on the level of user satisfaction with a service or initiative. Whatever the specific tools used, the aim of participatory monitoring is to provide mechanisms for people and organisations 'on the ground' to communicate their experiences about what is or isn't working well, so that improvements can be made. Participatory monitoring tools are a valuable real-time management resource; they can also be used as mechanisms for evaluation and accountability (see the next section).

Evaluation tools: Learning and accountability

From an anthropological perspective, evaluation is not just what happens at the end of a project. Through the processes of assessment, design and implementation, those who are involved in a development initiative learn about the contexts where they are working, the actors involved and the dynamics of change. Evaluation is not only about documenting outcomes, but continually learning about how to do development better.

Accountability has become a key concept in development work. Accountability refers to the need to understand what really happens when investments are made in development – and who really benefits. Earlier chapters observed that development initiatives do not necessarily succeed in making people better off; indeed, in some cases, they may do the opposite. Development professionals increasingly recognise the need for their work to be accountable for what it claims to achieve, and to provide evidence that it has made a difference. Funders, citizens and the would-be beneficiaries of development initiatives have a right to know how funds are spent, and who benefits from them.

Evaluation is the process of gathering data on the processes and outcomes of development initiatives, to inform learning and accountability. A number of practical tools are available that can be used to operationalise an anthropological approach to evaluation. An anthropological approach takes into account the perspectives of multiple actors and their experiences of change. It moves beyond a narrow focus on pre-defined outputs and outcomes to capture unexpected impacts and perspectives on the development process and its effectiveness.

Practice learning tools

A number of tools have been developed to facilitate *practice learning* – that is, practically focused reflection and learning about what works and what doesn't work, through a thoughtful analysis of practice experiences. In many development projects and programs, the focus on Monitoring and Evaluation (M&E) has now shifted to a focus on Monitoring, Evaluation, and Learning (MEL). MEL recognises the importance of embedding an ongoing process of practice learning into evaluation processes. The following tools can be used throughout the assessment, design and implementation of a development initiative to encourage continuous learning and improvement.

The action–learning cycle. The action–learning cycle is a classic and still useful way to visualise practice learning in general, and learning from development practice in particular. Originally articulated by Donald Schön, the action–learning cycle links action and reflection in a continual cycle of doing and learning.[15] In development work, the action–learning cycle can start with an action (such as a pilot project) or a reflection – that is, thinking about an existing situation or action. The continual dialogue between action and reflection is a simple way to think about how development practitioners and other development actors learn from what they do, and use what they have learned to inform their future work (see Figure 4.2).

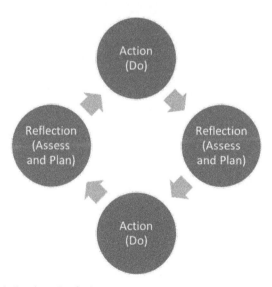

FIGURE 4.2 Basic Action–Learning Cycle

Particularly relevant to an anthropological approach to development work is the concept of *double loop learning*. Double loop learning is learning that actively reframes the problem itself.[16] Rather than learning something new that sits within established categories and frameworks, double loop learning questions existing categories. In development work, double loop learning happens when new knowledge from practice calls into question the established framing of development issues. Double loop learning makes it possible for development practitioners to take on board new knowledges, logics and ways of working that are outside current organisational comfort zones. It is thus closely related to the idea of reflexivity.

'Most significant change'. Various tools can be used within an action–learning cycle, at any stage of implementation or evaluation, to stimulate reflection upon what is working well and what can be improved. The 'most significant change' technique is a particularly useful tool for stimulating reflection and dialogue among development actors. The process asks development actors to consider, from their different perspectives and experiences, the most significant change they have observed in the implementation of an initiative.[17] A variant of most significant change is to explore the 'most significant moment' in a process. The focus on significant moments is useful to capture stand-out insights, experiences or lessons learned, in addition to specific stories of change. Both of these tools are compatible with an anthropological approach in that they explicitly value experiential knowledge, and recognise that different development actors, working from different social positions and development logics, will have different 'significant moments' and different perspectives on what comprises significant change.

Reality Check Approach (RCA). The Reality Check Approach (RCA) is an evaluation tool that development workers can use to ground their practice in a practical, experiential understanding of what life is like for the people they are working to help. In practice the RCA approach constitutes an immersion experience, in which professionals spend time living with a family 'in poverty', building a relationship with the family and their neighbours, and learning from them. This provides professionals with a first-hand experience of the development logics and experiences of their hosts, as well as providing qualitative insights on the actual impacts of development initiatives.[18]

Accountability tools

There is growing recognition that development initiatives need to be accountable not only to funders, but to their intended beneficiaries and other vulnerable groups. As a result, a number of tools have been developed to enable citizens and communities to directly track and evaluate the effectiveness of development work. Participatory accountability tools are very useful from an anthropological perspective because they provide a way for marginalised actors to share what they know and, potentially, to call more powerful development actors to account for poor results.

Community score cards. Community score cards and their variants (e.g. citizen report cards) are designed to allow community members to rate the outputs of an initiative or service according to a set of criteria.[19] Both the format and content of community score cards can vary considerably. They may literally be paper score cards, or they can be provided via electronic platforms. The score cards and their criteria may be designed by service providers to get input from user groups. Equally, the criteria may be designed by external funders or advocacy groups, or co-designed with community groups to encourage community-based evaluation. Finally, score cards may be directly designed by user groups or community organisations themselves.

As an evaluation tool, community score cards can be used on an ongoing basis to monitor the quality of a service or the overall satisfaction with an initiative or set of initiatives (e.g. local government services, health services), and to benchmark them against others. Score cards can also be used to provide evidence of positive impact or improvements over time. Community score cards are potentially powerful tools because they directly communicate on-the-ground assessments from the perspectives of the people on the receiving end of development initiatives. While many accountability-focused tools are concerned with monitoring expenditures and guarding against corruption, tools such as community score cards and other participatory monitoring tools, such as those described previously, can be used in evaluation to ensure that development initiatives are directly accountable to the actors they claim to help.

Tools for reflexive practice

Across all of the processes of assessment, design, implementation and evaluation, a key characteristic of an anthropological approach is reflexivity. Recall that reflexivity is the process of actively paying attention to the ideas and ways of working that guide our own practice, and in turn being open to the ideas and ways of working of other development actors. A few additional tools are included next which can be of great value in embedding reflexivity in development practice.

Immersions. Similar to the Reality Check Approach discussed earlier, immersions involve spending time on the ground in real social contexts, building relationships with local actors. Immersions are, however, more open-ended: they are not necessarily associated with a particular initiative or project stage; rather, they are simply 'occasions when professionals learn directly from encounters with poor and marginalised people by living with them and reflecting on the experience'.[20] The on-the-ground experience of immersion becomes an opportunity to learn from others, encouraging a reflexive approach that can reframe situations with attention to local contexts and the perspectives of the people who live in them.

Active listening. Active listening techniques are useful in engaging with other development actors. Rather than passively 'hearing' what someone says, active listening seeks to engage deeply with the content of what someone has to say, on its own terms. Rather than rapidly processing a statement according to our own categories and concerns (*she is being obstructive; he is missing the point*), an active listener seeks to understand, in detail, the points the speaker is trying to convey, in their own words. This encourages an openness to the speaker's perspective, and makes it easier to identify the knowledges, logics and ways of working that the speaker may be drawing on to make their points.

Strategic questioning. Strategic questioning techniques can be useful to move beyond accepted categories to 'unpack' deeper assumptions and issues. Rather than taking a situation or problem at face value, strategic questioning is a technique for exploring it in depth, either individually or in a group setting. Like peeling a metaphorical onion, questions like 'What does that actually mean?', 'How is that defined?', 'Why is that important?' and 'Who thinks so?' can peel away layers of assumptions and encourage critical reflection on a situation or problem in its social and cultural context.

Boundary spanning. Boundary spanners are people who regularly work across different organisations and sectors. Boundary spanning as a technique involves intentionally working across organisational or sectoral boundaries to create opportunities for communication and relationship-building. Boundary spanning is a way of encouraging learning and reflexivity by moving outside one's own comfort zone into different social and cultural settings. In development practice, boundary spanning can take a range of forms: from regular involvement in cross-sector initiatives, to mobilising professional and personal networks, to cultivating long-term relationships across boundaries. In whatever form, boundary spanning can help to grow an understanding of others' ways of working and identify opportunities for productive collaboration.

Working with anthropologists

The tools presented in this chapter can be used by development professionals to apply anthropological insights in day-to-day development practice. Yet there are also times when there is no substitute for a professionally trained anthropologist. Anthropologists have been applying their professional skills to development situations for decades: as consultants, researchers or staff members of development organisations. Depending on the initiative and the skills required, it may make sense to work with a professional anthropologist.

Anthropologists can support development initiatives in a range of ways.

As topic experts. Anthropologists who specialise in particular domains (such as environmental anthropologists, economic anthropologists and political anthropologists) can offer in-depth expertise on particular topics (such as forest management, microenterprise development or local government). Their expertise is informed not only by scholarly insights, but by their real-world experiences learning from people in the places where they have studied. Employing anthropologists as topic experts thus delivers the best of both worlds: specialised scholarly knowledge, and understanding of the other kinds of on-the-ground knowledge that can be deeply relevant to development work.

As area experts. Anthropologists often have one or more geographic areas of expertise. In some cases they have spent years living in particular places to conduct research with particular groups of people. This means that they have in-depth and often long-term knowledge of a certain area and the social groups who live there: the history, the language, the social roles, how things work and what has been tried before. Where an organisation has limited knowledge of a particular context or cultural group, it can be enormously beneficial to hire an anthropologist as an area expert to provide orientation, advice, background information, and potentially even introductions to key people and groups.

As methodology experts. Anthropologists are skilled in field methodologies and methods that are designed to capture an in-depth view of local contexts and the perspectives of local actors. Participant observation, in-depth ethnographic interviewing and artefact elicitation are only a few of the methods that trained anthropologists may use to gather rich data about local contexts. Further, some anthropologists have developed or adapted practice-focused and participatory methods. As methodology experts, anthropologists offer creative approaches to understanding development contexts 'on the ground' and 'from within'.

As brokers and mediators. Anthropologists tend to have strong skills as cultural brokers, with the ability to work with people and groups across social and cultural boundaries. Their ability to understand and express the viewpoints and ways of working of very different groups equips them to effectively facilitate cross-cultural communication and relationship building. Anthropologists can broker external engagement with unfamiliar and 'hard-to-reach' groups, as well as helping marginalised groups to advocate for their interests and engage with more powerful social actors.

As critical friends. There is an argument that every project team should include an anthropologist. In addition to the skills they may bring in other areas, anthropologists have, as part of their training, deep skills in reflexivity and cross-boundary thinking. Anthropologists are trained to work across cultural contexts; this brings with it the ability to question established categories and seek out alternative perspectives. Anthropologists as critical friends will question, challenge, and suggest new ways of approaching old issues. Simply by the way they see things, anthropologists can catalyse creative problem-solving.

Finally, with the advent of new technologies, anthropological expertise is often closer than you think. A recent innovation in the use of anthropology in development work is the *online digital knowledge platform*. Online digital platforms can directly link anthropologists and development professionals around specific areas of practical need. The Ebola Response Anthropology Platform (ERAP), for instance, was developed to provide a direct channel of communication, data and advice to aid workers about managing the socio-political and cultural dimensions of the Ebola outbreak. This digital platform includes practical online briefings, guides and the facility to ask a 'rapid response question' to members of a global network of anthropologists.[21]

Summary: Tools and their uses

This chapter has presented a cross-section of practical tools which development professionals can use to apply an anthropological approach in their day-to-day practice. These tools can be used within the constraints of standard assessment, design, implementation and evaluation processes to reveal otherwise-hidden aspects of the development landscape. They can uncover key characteristics of the social context, give voice and authority to a wider range of actors, challenge stereotypes and assumptions, and encourage 'double loop' learning and reality-checking that reframe practice with attention to the perspectives of marginalised groups.

The tools presented in this chapter are not, however, 'answers' to the challenges of development work. Like all tools, they can be employed well or badly. The same tools that can be used to identify and listen to diverse actors can be used to reinforce stereotypes and assumptions about those actors. The same tools that provide mechanisms for more participatory, inclusive or empowering development processes, can easily be used to exclude or silence particular groups, or to reinforce existing relationships of privilege and dis-advantage. While each of these tools can be used to take an anthropological approach, they are not necessarily always used in this way.

This is the paradox of practical tools: moving from theory to practice requires tools and approaches that are very concrete and can be 'rolled out' on the ground in real practice situations. Yet these concrete, practical contexts are also eminently social. Tools are used by people; and the actions of actors matter. Tools will work differently depending on who is using them.

An anthropological approach, therefore, can never be just about tools and methods. It ultimately requires attention to the actors involved, as well as to one's own social position as an actor on a complex development landscape. For development professionals, this means an orientation toward reflexive practice: a way of working that continually interrogates one's own role, social position, assumptions and aims, and how to relate ethically and respectfully to other development actors – especially those we seek to help. These ideas will be explored further in Chapter 5.

Notes

1 The idea of the Overton window is useful here, as is Haugaard's concept of 'infelicitous' action (see Eyben *et al.* 2006). Both concepts emphasise that ideas and actions that are too far outside current comfort zones are deemed unacceptable; there is only a limited 'window' of institutional options for change that people will accept.

2 In development work, initiatives start 'now' and look forward to an improved future. This focus on the present and future is logical and optimistic, but it has a disturbing corollary: the tendency for development work to forget the past. Anthropologist David Lewis (2009) has identified that development work tends to exist in a 'perpetual present', and forget that current situations have resulted from past actions. He notes that 'project documents may over-focus on the present at the expense of the wider context, politics, and history' (pp.33–34). An anthropologically informed assessment will seek to understand current situations with an eye to the institutions and power relations that have created those situations over time.

3 For instance, Sondra Hausner (2006, p.331) notes that: 'anthropology is best used *before* a development project begins. . . to help define the contours of a problem and the way informants conceive of possible solutions' (italics in original).

4 Pottier (1993, p.3).

5 Stakeholder mapping can also be conducted in workshop environments; the mapping of actors and their relationships is often a component of other approaches such as Participatory Rural Appraisal (PRA) and Participatory Impact Pathways Analysis (PIPA).

6 Chambers (1983).

7 See, for instance, Burns and Worsley (2015) who describe three practical approaches to systemic enquiry.

8 Chambers (1994) has written on the origins of the PRA approach; a number of handbooks of PRA methods are available.

9 A range of resources are available on participatory visual methods; see, for instance, www.participatorymethods.org.

10 See www.participatorymethods.org, Brown *et al.* (2005) and Chambers (2002).

11 For more detailed information on PIPA see http://pipamethodology.pbworks.com/ w/page/70283575/Home%20Page. For an example of an organisation that uses this approach regularly in their work, see https://steps-centre.org/pathways-methods-vignettes/ methods-vignettes-participatory-impact-pathways-analysis-pipa/.

12 See e.g. http://participatorymethods.org/glossary/participatory-budgeting.

13 The Partnership Analysis Framework (Eversole 2015, p.82) can be used to analyse the logics and power relations in development partnerships.

14 See http://collaborationforimpact.com/collective-impact/.

15 See e.g. Schön (1984).

16 The concepts of single loop and double loop learning were developed by Argyris and Schön (1974, 1978).

17 For more information on the 'most significant change' technique, see http://better evaluation.org/sites/default/files/EA_PM%26E_toolkit_MSC_manual_for_publication. pdf and http://mande.co.uk/docs/MSCGuide.pdf.

18 See https://bond.org.uk/resources/effective-development-group-short-introduction-reality-check-approach.

19 See the Participatory Methods website: http://participatorymethods.org.

20 Irvine *et al.* (2004, p.3). For more information on immersions, see http://eldis.org/vfile/
    upload/1/document/0903/IMMERSIONS2.pdf.
21 ERAP is online at http://ebola-anthropology.net/; this initiative won the UK Economic
    and Social Research Council's 2016 international impact prize. Melissa Leach, one of the
    architects of the platform, has described the process of establishing this global platform
    for engaged anthropology, and its contribution toward encouraging health systems to
    become more effective by working 'in more community-realistic and embedded ways'
    (2015, p.7).

5

Anthropological responses to development challenges

Development problems are remarkably long-lived. While an average career in development work may span forty or fifty years, most of the problems we grapple with pre-date that considerably. Many stubbornly refuse solution, even after mountains of money, skills, expertise, time, people, energy and political will are employed in a concerted effort to *do something* about them. This chapter explores how an anthropological approach can help us to make sense of big-picture problems and challenges of development work.

Around the world, enormous effort and resources continue to be poured into development initiatives, at every scale from the local neighbourhood or municipality to major multilateral programs. Development professionals work to create change at a wide range of scales: from improving local economic opportunities for a rural council, to spearheading industry development policy for a national government, to leading a women's empowerment program globally. In every case, development professionals work to make a difference. Yet all face the core challenge of *development effectiveness*. Why do so many development efforts fail to create real change?

Poverty is no longer understood to be a passing stage that countries pass through and graduate out of; it is now recognised as a persisting, global problem across 'developed' and 'developing' countries alike. Poverty is recognised as multidimensional – not just about income, but also things like access to services, status and political voice. Yet significant poverty persists despite decades of anti-poverty initiatives. Some people in some places have improved their situations, but others are the same, or even worse off. Inequality endures, and in many cases increases.

Three decades ago, ideas like *participatory development* and *sustainable development* promised to revolutionise development work. Participatory development attempted to place the needs, resources and aspirations of so-called poor communities at the centre of the development agenda. Sustainable development challenged the prevailing idea of development-as-economic-growth, and proposed new ways of thinking about development embedded in ecological and social contexts. Both sets of ideas recognised that effective development work must be contextualised, centred on people and their lives in real places.

Thirty years on, the language of participatory and sustainable development has entered the mainstream. Yet development work itself has changed very little. The failure of top-down, one-size-fits-all development efforts has been well documented; the need for contextualised, community-based approaches to solving problems is now broadly acknowledged. Yet 'top-down development' is alive and well. Indeed, with the growing trend toward high-level coordination and common policy frameworks, development practice is arguably more top-down than it was two or three decades ago.[1] People-centred social change seems further away than ever.

Development professionals in the twenty-first century are facing a set of entrenched challenges which the development sector itself is ill-equipped to solve. Overseas aid budgets are cut and criticised; domestic programs for disadvantaged communities are cut and maligned. Often it is the participatory, people-centred, context-sensitive programs and projects that are the first to go. Meanwhile, development professionals struggle to face hard questions: *Why do our projects, programs and policies so seldom work as well as we had hoped? Who has really benefitted from our efforts? In the end, are we making a difference?*

It is possible to do development work without asking these questions. One simply continues to deliver development projects, programs and policies within the traditional development practice framework as visualised in Figure 3.1 – problems are identified, target groups defined and solutions crafted, in line with currently in-vogue policy beliefs about change. So long as the initiatives line up with the high-level policy frameworks, so long as they are directed to the politically popular target groups, and so long as they are evaluated according to the criteria valued by the policy makers, they look successful.

Of course, looking 'successful' at desk level and creating a real change on the ground can be very different things. Just as building a school does not automatically create education, managing a well-performing development program does not necessarily create development. For this reason, development problems can persist and even worsen, even in the face of decades of initiatives to solve them. Indeed, the illusion of 'successful' development can be actively created among those with an interest in maintaining programs or justifying policies. Anthropologists have documented that even members of target groups may be willing to support the appearance of development success, if they see this as a way to secure future resources or influence from powerful outsiders.[2]

This does not mean that development never works. Many initiatives do make a significant impact on poverty and disadvantage – and many more could. Development efforts can be highly effective, but there are sound institutional reasons why ineffectiveness is common and persistent. Development work has structural characteristics that regularly insulate it from a need to question its role too closely – or to actually notice failure. Framing development within developer-defined categories creates decision-making processes that are largely disconnected from reality on the ground. This allows initiatives to self-perpetuate largely independent of what is actually going on in the places and communities they are intended to benefit.

Decades of anthropological research have documented why many development initiatives fail: because they don't understand the contexts they are attempting to influence, because they ignore or stereotype local actors, and because they fail to recognise actors' diverse social positions, their knowledges of local contexts, and their functional – if unfamiliar – social and economic institutions. Development initiatives enter living social systems with the aim to create change, yet often with limited or no understanding of what is already there on the ground. They perpetuate development ignorance. Then, when things go wrong, it is nearly always those with fewer resources and safety nets who suffer most.

This chapter looks to the future of development work in light of today's key challenges. It explores how an anthropological approach can help development professionals to tackle, reflexively, the hard questions of development – questions like: 'Are we making things better?' and 'Are we making things worse?' It uses an anthropological approach to unpack some of the 'grand challenges' facing development practice in the twenty-first century:

1 The challenge of *development effectiveness*, or overcoming waste and failure;

2 The challenge of *fighting poverty*, or really making a positive difference for 'the poor';

3 The challenge of *participation*, or ensuring less powerful groups really do get a say; and

4 The challenge of *sustainability*, or achieving triple-bottom-line benefits over the long term.

These four 'grand challenges' cannot be solved within the framework of development business-as-usual. Decades of trying have only proven the point. Yet neither are these challenges as insurmountable as they may sometimes appear. An anthropological approach to development reframes effectiveness, poverty, participation and sustainability, and suggests new strategies for change.

A recipe for effectiveness?

There was a time when the effectiveness of development work was simply assumed. Financial investment and technical know-how, delivered when and where they were needed, would – it was generally agreed – create development.[3] It seemed an easy recipe for success. Yet this optimistic prognosis did not hold true in practice. Many development initiatives over the years have invested enormous money, time and expertise to make no impact at all. Even worse, there is abundant evidence that some large investments aimed at creating positive change have actually made things worse: depleting resources, endangering livelihoods, exacerbating inequalities.

This depressing cocktail of waste and failure has led to a growing disenchantment with development – reflected in slashed foreign aid budgets, diminished domestic development programs and a growing obsession with value for money and evidence-based decision making, as decision makers struggle to document that investment in social or economic development is money well spent. Even development workers can become disenchanted when things regularly do not turn out as planned: in one study in India, public officials charged with delivering a state development program frankly expressed their doubts that their highly politicised program would ever yield results, and expressed a desire for 'real development' instead.[4]

'Real development' is *effective* development; initiatives that make a positive difference. The core question of development effectiveness is fast becoming a central grand challenge for development practice in the twenty-first century. How can we ensure that our development efforts make a real, positive impact? And can anthropology help?

An anthropological approach suggests that there is no one recipe for effectiveness. At the same time, thinking anthropologically about development can help us to approach the challenge of development effectiveness in a more fruitful way than current efforts do. By focusing on how our current institutional ways of working either challenge or perpetuate development ignorance, an anthropological approach can show us where our efforts to make a positive difference most often go wrong, and what is required in order to do things differently.

Currently, policy makers tend to take one of two approaches to grappling with the challenge of development effectiveness. On the one hand, they seek out 'market-based' solutions to delivering development goods, in the hope that market-based strategies will increase efficiency, responsiveness and ultimately value for money. On the other hand, they demand more evidence and rigorous evaluation processes to guide their development investment decisions. Both of these are logical responses to the need to ensure development effectiveness in an environment of incomplete information and limited resources. Yet neither approach actually addresses the root problem of why most initiatives work poorly. Further, these responses bring additional problems of their own.

Market-based solutions are popular because they promise to expand the pool of available resources for development work – for instance, through attracting private-sector investment or generating income through social enterprise.

Market mechanisms also promise efficiency and effectiveness, since initiatives that fail to meet a real need will simply not get off the ground. Market-based solutions underpin a whole range of contemporary development practice initiatives, from public–private partnerships to the commercialisation of microfinance.

Unfortunately, while market-based solutions can indeed be cost-effective and market-responsive, they can also be exclusive and inequitable. Market-based models may easily price important public goods like water and education out of reach of many people, or fail to meet needs that are unprofitable to serve. Still, policy makers cling to the illusion that market mechanisms can solve the problem of development effectiveness. Rather than solve it, however, they simply hide the problem from view.

Evidence-based approaches to development effectiveness are popular because they promise to show which development investments work well and which do not. It has been observed that in development work 'performance measurement and audit have now become key preoccupations'.[5] Evaluations of rigorous impact measures are increasingly used to make a case for development effectiveness. Evidence of results can be used to justify investment, while also enabling decision makers to 'weed out' ineffective or poorly performing initiatives.

While it is difficult to argue with the virtue of evaluating results, a rigorous evaluation does not necessarily say a great deal about effectiveness per se. It ultimately depends on what is being measured. If the results that are being evaluated are not things that really matter on the ground, or if other important things have changed but are not measured, then the evaluation data says little about whether or not a development initiative has actually made a difference. Indeed, tightly structured, quantitative evaluation processes of the type most favoured by policy makers can easily miss unintended consequences, both positive and negative.[6]

Development effectiveness thus remains, for all intents and purposes, an unresolved issue for development work in the twenty-first century. Across both poor and wealthy countries, neither market-based nor results-based approaches are able to provide a convincing response to the problem of development effectiveness. Each of these strategies continues to define 'effectiveness' using the development logics and metrics that make sense to policy makers, while ignoring the logics and metrics of the people they are aiming to help.

An anthropological approach suggests a very different strategy for grappling with the challenge of development effectiveness. It requires looking to the root of *ineffectiveness*: in short, development ignorance. From an anthropological perspective, development effectiveness will never be achieved from within the same framework that creates, and continually re-creates, ineffectiveness.

Throughout this book, we have seen that development efforts often fail because those who make development decisions fail to understand key aspects of the development landscapes on which they are trying to create change. They do not take on board local knowledge or notice how local institutions work. As a result they often demonstrate a basic ignorance of a number of things that matter deeply for development effectiveness. For instance:

1   Projects, programs or policies often fail to take on board local knowledge about the specific *physical* characteristics of the environments where they are expected to play out, and which may be very different than outsiders assume: for instance, terrain may be rugged, roads impassable, power or communications infrastructure inadequate, or climatic variables very different than expected.

2   Projects, programs or policies often fail to accommodate the specific *social* contexts of the environments where they are expected to play out. They forget that local people have diverse interests and social positions. They overlook that local people are savvy; those who can will mobilise their resources and influence to channel change in the direction they prefer. And they fail to recognise that people are situated within particular practical and social constraints that affect what they can or cannot do: as a woman, a traditional leader, a young person, a single parent and so forth.

3   Further, many development initiatives are largely ignorant of the *historical* and *institutional* factors that continue to influence current attitudes and logics. Development initiatives tend to start their planning with the present; but many institutions that affect people's room to manoeuvre have deep roots in the past. Thus, past experiences of colonisation can persist in present-day power differences. And historically marginalised groups can find that they are still marginalised and stereotyped into the present.

4   Finally, many development efforts fail to appreciate the *interconnectedness of different domains* in a particular place: for instance, how land-use choices can impact environmental and human health, or how political imperatives influence economic decision making. Most development initiatives are designed and implemented in silos: health here, education there, economic development over there. Yet on the ground, in people's lives, these domains are interrelated. Change in one domain affects, and is affected by, circumstances in another – often in unexpected ways.

Most development failures are unsurprising; they could have been quickly predicted by anyone with any knowledge of the context. The flip side of this is, of course, that most failures are eminently preventable. The key is to start, not with an abstract *problem*, as per Figure 3.1, but with a concrete *context*, as per Figure 3.2. Knowledge of context is fundamental knowledge for development effectiveness. To meet the challenge of development effectiveness, it is necessary to confront the problem of development ignorance.

This is a straightforward proposition; yet it has deep implications for how development work is done. When we consider how the majority of contemporary development work is structured, development ignorance is understandable. Development workers are commonly isolated from real-world contexts – making their policy and program decisions from head offices of agencies or governments, located far from the people they aim to help. Glynn Cochrane observes, for instance, that 'seventy to ninety percent of the staff of global aid agencies and civil society organisations spend most of their working careers in large offices in New York, Washington, DC, Geneva, London, and Paris'.[7] Even within the same country, significant social and physical distance often separates developers from developees. Tom Gabriel has observed that those who manage development programs are 'usually extremely remote from the living conditions of their clients' and so lack 'fundamental knowledge' needed for good development practice.[8]

The question of development effectiveness is ultimately a question of who makes the decisions. That is, it is a question of governance. *Governance* is the structure of decision making; it is about who makes decisions and how, whether for a single organisation (e.g. the governance practices of a board) or for society as a whole. Currently, governance – of countries, cities or development organisations – tends to be centralised: senior people, located in head offices, make most of the decisions that matter. Yet these people are generally located far away from the contexts where their decisions 'hit the ground', and they lack experiential knowledge of what these decisions mean for the people who live in these contexts. The physical and social isolation of decision makers breeds development ignorance.

The challenge of development effectiveness is, in the end, an opportunity: a chance to look more closely at the role that governance arrangements can play in making development work more effective. Decentralised decision making, multi-stakeholder processes and participatory governance offer mechanisms to challenge development ignorance. In particular, governance processes that include historically marginalised actors directly in decision making are more likely to learn from their own knowledges and institutions, and identify solutions that successfully meet their needs.

Poverty as a verb

Poverty is a global 'grand challenge' – and one that the development sector grapples with on a daily basis. While development work can have any number of political agendas, the goal *to fight poverty* captures the public and policy imagination. From movements to 'Make Poverty History', to the number one Sustainable Development Goal (SDG), 'Eradicate extreme poverty', fighting poverty is a well-regarded project, supported by governments, philanthropists, multilateral agencies, NGOs, socially responsible companies, and community organisations around the world. Why then, does it appear that poverty is winning?

An anthropological approach can provide some helpful answers. Specifically, lessons learned from anthropology suggest that in trying to attack, fight, reduce or eradicate something called poverty, we are getting the focus wrong. Poverty is not a homogeneous thing – a noun – that can be identified and then eradicated. Rather, poverty can better be understood as a process or set of actions – a verb – through which particular people and groups are systematically and predictably deprived of resources and influence.

Anthropologists reframe poverty with attention to the social contexts where poverty happens. They focus on the actors in these contexts – people and organisations who are diverse, savvy and situated in ways that affect their room to manoeuvre. In *Festival Elephants and the Myth of Global Poverty*, anthropologist Glynn Cochrane argues that development agencies' vision of poverty is based on 'an image of billions of faceless and voiceless poor people', rather than an understanding that 'poor people everywhere, from Tijuana to Tahiti, are poor for different reasons and experience poverty in different ways'.[9] For anthropologists, 'the poor' are no homogeneous group.

Nor is poverty something that poor people 'have'. Rather, when anthropologists write about poverty, they emphasise that poverty can be understood by looking at what actors *do*. Poverty, Maia Green writes, 'is not a "thing" to be attacked, but the outcome of specific social relations that require investigation and transformation'.[10] Katy Gardner and David Lewis similarly explain that, 'Poverty is, first and foremost, a social relationship, the result of inequality, marginalisation and disempowerment.'[11] Poverty is the action of depriving; what some actors do to others.

In development work we concern ourselves with measuring, quantifying, analysing, attacking and ultimately eradicating a thing called poverty. The problem with seeing poverty as a noun, a thing-in-itself, is that it is then assumed to be a concrete, fixed state, rather than a fluid process of deprivation. Poverty as a noun provides a way of comparing people or ascribing an identity to them ('the poor' or 'poor communities'); yet these may bear little relation to how these people see themselves. Poverty as a set of metrics or attributes describes what is lacking – income, assets, education or political voice – yet gives few clues as to why someone lacks them.

By contrast, poverty as a verb asks why resources and influence are lacking. What processes block resources and influence for certain actors, and so are poverty-creating? An anthropological understanding of poverty reveals that it is not an identity or social category, but rather a process that people experience. Further, this process often has deep historical and institutional roots. In poverty-as-a-verb, some people and organisations systematically prevent others from accessing resources and influence. People of one race, gender or family group block access for others. The institutions of some groups leave others little or no room to manoeuvre. Poverty *happens* to people through their interactions with other actors and their institutions. It is not something that they have, like a skin colour or birth language; it is what they experience.

Because poverty is an active process rather than a state of being, it can be changed. Seeing poverty as a verb recognises that social dynamics create and recreate situations of lack. This reframing provides a concrete direction for anti-poverty action. To eradicate 'poverty' it is necessary to eradicate poverty-creating, the process. This requires doing things differently. It requires asking: How do current ways of working reinforce poverty-creation? How could changing our ways of working change the actions of poverty?

With this reframing, the focus of anti-poverty work widens: from the traditional focus on target groups of the poor, to the broader landscape of actors whose actions and interactions create poverty-as-a-verb. The corollary to this is that most efforts to solve poverty are starting in the wrong place: with 'poor people' and what they objectively lack (poverty as a noun) rather than by trying to understand the processes that create the lack (poverty as a verb).

Most anti-poverty initiatives start from the assumption that 'poor people' are the problem, and thus the ones who need to change: by becoming more productive, educated, healthy, entrepreneurial, empowered and so forth. Looking up-close at local contexts, as anthropologists do, reveals that it is seldom 'poor people' who are the problem. Generally, people are already trying to do what they can to change their situation, within the limited room to manoeuvre available to them. Yet the institutions they work within, and the other actors they work with, severely limit what they can do and the resources they can access. Despite their best efforts, their world in the end is 'full of rotten choices'.[12]

'The poor' themselves seldom create poverty. Targeting poor people with programs designed to change them – to make them more educated, productive, politically engaged, etc. – often overlooks their own efforts to improve their circumstances. Anti-poverty initiatives frequently layer on new workloads and risks to people's existing commitments without asking the obvious questions: Why aren't people's current efforts to improve their circumstances working? Where do the blockages actually lie?

Answering these questions requires looking at the institutional contexts where the people who experience poverty processes live and work. Poverty processes are social and cultural processes, and they manifest in people's and organisations' accepted ways of working – that is, in their institutions. Often, these institutions have long histories, and discriminatory or disadvantageous ways of working have been practised for so long that they have become an unconscious norm. Poverty can be seen in action in practices of institutionalised racism, corruption, injustice, bias, and in the myriad ways that relationships can systematically perpetuate disadvantage.

The processes of poverty-as-a-verb happen daily and in often predictable ways: in rules that exclude, in actions that belittle, in experiences of disdain and disrespect, in deprivation and isolation. In every social context, established ways of working reinforce the message that some people and groups are valued and privileged, others not; that some are simply meant to have resources and influence, others are not. Eradicating poverty thus requires reflexivity: it requires people and organisations of all kinds to be willing to look beyond their own framings of business-as-usual and notice poverty happening – in the interactions among people and organisations that actively create disadvantage. Paying attention to how these processes work provides the best clues about how they can change.

From participation to recognition

Despite a tendency to portray people as the passive 'targets' of development interventions, development professionals generally recognise that in principle, people should be active participants. The idea of *participatory development* proposes that development should not be done 'to' people or 'for' them, but 'with' them. It aims for people to participate actively in the process of designing and implementing development initiatives to meet their needs.

Development professionals spend a great deal of time considering how to encourage greater participation, particularly by those who are disadvantaged and 'hard-to-reach'. Nevertheless, a number of practical issues have plagued efforts to make development practice more participatory. Despite decades of effort, very little has changed.

An anthropological approach can explain why participation remains an elusive goal, and what can be done in response. Anthropology's attention to development actors highlights that participatory development processes are nearly always pre-framed around the interests of developers. Whether the developers in question are in a municipal council, a government department or an international NGO, their definitions of problems and their preferred institutional turf dominate the relationship with would-be participants.

An anthropological approach reframes the challenge of participatory development to consider the perspectives of other development actors. Why would they choose, or not choose, to participate? What do these processes look like from their perspectives? This approach to participation provides a new view on old questions, such as why it is difficult to attract and retain participants, and why it is difficult to convert participants' insights into real change.

For instance, a common issue in development practice is that it is difficult to engage people in participatory processes. 'Participation fatigue' is common; people are not interested in participating, or they start out enthusiastic and then fall away. Focusing on the would-be participants as development actors reminds us that time and effort are required for participation in development efforts – usually unpaid time, and effort by people who have few resources to start with. Hosting and orienting project staff or government representatives, or participating in assessment or planning meetings, all involve time, energy and opportunity costs – this time and energy could have been spent in other ways. A recent study calculated some of the costs of participation for remote Indigenous Australian communities; one community with 240 adults hosted 1959 visitor-days of public- and private-sector representatives in a single year. Most of these were consultative visits, with few opportunities for real decision making.[13]

Another factor influencing actors' decisions about whether or not to participate is their assessment of the likelihood that the time and effort invested will yield any useful results. Community consultations, for instance, promise to provide a route for local knowledge and priorities to inform policy; yet ideas from ground level rarely seem to reach the ears of policy makers. Actors have memories; they recall what happened the last time they were asked to invest time and effort in a process, after which nothing seemed to change. For instance, consultations in rural Australia often earn the epithet, 'just another talk fest'. Everyone has talked, but conversation has not led to action. People remember: we send ideas up – but nothing ever comes back.[14]

Many efforts at participatory development are asking development actors to participate in other people's agendas – and to use their own resources to do so. The 'usual suspects' may show up at a meeting because they believe they can benefit in some way, or because they value the relationship with the organisation that is convening the meeting. In some cases they may have employers who will pay for their time, or they see project meetings as a useful social outlet. But for other actors who are deemed 'hard-to-reach', they do not see that the benefits of participation would outweigh the costs. They are generally aware that participation opportunities will be pre-framed by the interests and agendas of other actors, and may see these as low priority, irrelevant to their lives or unlikely to change in any case.

The original promise of participatory development was to change how development work is done. In practice, however, participatory tools and techniques usually end up positioned as cosmetic add-ons to mainstream ways of working. Where developers attempt to integrate a participatory approach, participation can quickly highlight the tensions and fault lines in mainstream development practice. For instance, participatory processes require time to build trust and dialogue with participants; yet development project time-frames are sharply limited and often need answers yesterday. Further, participatory processes regularly throw up issues that are 'out of scope' for organisational portfolios; they must either be refocused into something the organisation can actually do, or be ignored. These tensions between mainstream practice and participatory approaches are common.

An anthropological response to the challenge of participation locates the problem of participation in deeply held ideas about the roles of 'developers' and 'developees' in development work. Developers – individuals and organisations – see themselves at the centre of the process, responsible for defining and managing a carefully crafted change process – albeit with some input from participants. Yet these participants are also actors who have their own logics and processes, and their own agendas for change. The concept of *participation* has strong overtones of joining other people's agendas. To revisit people's right to participate in the decisions that affect them, a more useful concept is *recognition*.

Recognition involves reframing the role of would-be participants: not as developees, beneficiaries or participants in someone else's process, but as development actors in their own right, who have a right to have a say in the direction and nature of change. Borrowed from the language of international Indigenous rights movements, the idea of *recognition* moves away from the experience of *participation in someone else's agenda* to the idea of *participation in relationships on one's own terms*. Recognition is thus a political position, an acknowledgement that the communities and organisations of participants have rights, and an active role in defining issues and solutions on their own terms.

The idea of recognition challenges development professionals to recognise the validity of who other development actors *are* and what they *know* and *do* – that is, their own knowledges and institutions. The first step toward recognition is simple but profound: for development professionals to recognise the legitimacy of other forms of knowledge and ways of working. Development professionals tend to define problems and solutions in a way that excludes and dominates other knowledge traditions.[15] For instance, it has been observed that many of the commonly used words in international development work come from English, and often remain unchanged in local languages as foreign loan words.[16] Successful community organisations in any country are those who have learned to employ policy language and buzzwords fluently when they speak to the press, prepare funding applications or lobby for support. Meanwhile, those who are not able or willing to express their perspectives in this way, usually remain unheard.

An anthropological approach suggests that resolving the challenge of participation requires moving from surface 'participation' to respectful 'recognition'. Currently, development action still takes place within the institutional context of the developers, and in cultivated isolation from the larger development landscape. Participating in development initiatives requires acquiring fluency in developers' terminology and ways of working. Recognition flips the terms of engagement, suggesting that rather than engaging participants in their efforts, development professionals might more productively be engaged as participants in processes designed and led by the actors they seek to help.

Recognition emphasises the legitimacy of others' ways of knowing and working, as well as an awareness of what we ourselves know and do. It thus requires a conscious position of reflexivity – looking past our own personal and organisational framing of issues to see and engage with other actors on their own terms. Reflexivity involves a deep awareness of our own social positioning; the idea that 'how I personally understand and act in the world is shaped by the interplay of history, power and relationships'.[17] This in turn opens up an awareness of the positions and perspectives of others.[18] As Michael Drinkwater put it, a useful strategy for development practice is 'deliberately attempting to shift or alter our perspective until we comprehend the world from another's position'.[19]

Without recognition, even the best-meaning attempts at participation will simply reproduce old relationships and power structures. Developers or powerful interest groups will set the agenda; participants will comply or opt out. Old ways of working easily persist in new clothes. From an anthropological perspective, the challenge of participation is real; it is also an enormous opportunity. Moving from participation to recognition multiplies the perspectives we can see, the voices we can hear and the options that are available for change. In development practice, it opens up the potential for innovative, co-created solutions.

The challenge of sustainability

With the launch of the SDGs in 2015, the language of sustainability moved to the centre of global development policy. Sustainable development is perhaps *the* grand challenge for development work in the twenty-first century. Yet it is a challenge which is difficult to define: sustainable development is not a specific process or goal; rather, it is a philosophical commitment to a certain kind of development. To append the level 'sustainable' indicates a commitment to development that is both *lasting* (rather than of short-lived benefit) and *multidimensional* (rather than focused on economic outcomes alone). The language of sustainable development thus starts to articulate the core challenge facing development practice: how to create lasting and multidimensional positive outcomes.

Despite a global commitment to sustainable development, current development practice is poorly equipped to deliver both lasting and multidimensional change. An anthropological approach can help explain why, and how to respond. First, it is difficult for development solutions to be *lasting* when they fail to engage with people's own institutions, as these are the most logical vehicles for sustaining change over the long term. Next, it is difficult for solutions to be *multidimensional* when they persist in tackling development issues in silos, rather than recognising the interrelationships among different aspects of people's lives that are observable in local contexts. An anthropological approach suggests that the challenge of sustainable development is essentially the challenge to re-situate our work with attention to people and context.

In recent decades, sustainable development has articulated a persistent challenge to development practice to do change-making better. Originally, *sustainable* development was articulated as an environmental counterpoint to the dominant economic growth paradigm in development work. It served to remind decision makers that environmental resources are finite. Then, *responsibility to future generations* became a key tenet of sustainable development.[20] The focus on an intergenerational time scale was accompanied by growing attention to the social aspects of sustainability. The term *sustainable development* began to be used to posit that change processes have multidimensional, *triple-bottom-line* impacts: that is, social, economic and environmental.

Many of these ideas about sustainable development – such as the limits to economic growth, or the need to take social and environmental impacts into account – were considered somewhat radical when first proposed, but have now largely entered the development mainstream. Practically everyone now acknowledges the need for development to produce multidimensional, triple-bottom-line outcomes: improving the economy while damaging the environment or communities is not positive change. Equally, practically every-one now acknowledges, at least in principle, that positive change is a long-term proposition: change that does not benefit future generations is not sustainable change.

'Sustainable development' has become a way to make certain points about what development needs to be, and to gain widespread agreement about them. The SDGs include numerous aspirations for *positive and lasting social, economic and environmental change*; they articulate a shared aspiration and commitment globally. The challenge of sustainability remains silent, however, about how this big-picture shift toward lasting and multidimensional change is actually to be achieved.

This sustainability challenge cuts to the heart of development effectiveness. If our work is not sustainable, then it is not effective. Yet short-termism is rampant in development work: governments think in political cycles, and projects and programs end with the conclusion of funding. Positive change can be achieved to some extent, but change that *lasts* beyond the life of a project or program is much more elusive. Further, it is common for development initiatives to succeed in one dimension of sustainability (e.g. economic growth or environmental rehabilitation) and to fail in others. *Multidimensional* positive change is very difficult to achieve.

For anthropologists, the observation that development initiatives are often unsustainable is unsurprising. Context matters deeply to sustainability; yet mainstream development work regularly defines problems, target groups and solutions in isolation from local contexts. Time limits and sectoral bound-aries are built into the institutional structure of development organisations.

Development initiatives hit the ground in complex social contexts with multiple actors, yet persist in portraying change as something tightly framed and controllable. Developers draw upon their own institutional apparatus for change-making and forget that they are only one of many actors and activities on a diverse development landscape.

The results are eminently predictable. Developers often miss the inter-connections among economic, environmental and social processes on the ground. They focus on specific problems and so overlook how a change in one area can have flow-on effects – often unintentional – in others. Programs may succeed in raising incomes, but at the expense of increasing women's already high workloads; policies may succeed in providing needed infrastructure, but at the expense of fragile, livelihood-supporting ecosystems. Rather than see the connections among economic, social and environmental activities, development decision makers persist in a siloed approach that makes triple-bottom-line impacts nearly impossible to see – let alone to achieve.

Further, developers often tend to 'layer on' new and unsustainable project- and program-based ways of working into local contexts, rather than building upon what is already there. Instead of starting with how things currently work in a particular context, developers tend to set up new institutional structures that make sense to them. It has been observed that 'not even the best-intentioned NGOs are exempt from the tendency of the Development Industry to ignore, misinterpret, displace, supplant, or undermine the capacities that people already have'.[21] Developers' ways of working may duplicate, sideline or even undermine how people prefer to work, and their processes that are already working well. And institutions imposed from elsewhere are unlikely to last.

An anthropological response to the sustainability challenge recognises that it is not possible to create multidimensional, lasting development impacts in isolation from local contexts. Sustainability cannot be just an add-on to development-as-usual. If a development initiative is ignorant of contextual factors, if it conflicts with other local ways of working, or if it throws local social or environmental systems out of balance, it is unlikely to outlive the presence of its proponents. Further, those who instigate the change may never even see the effects that reverberate well beyond their own project, program or policy frame. At the same time, many organisational actors are already present on the development landscape with in-depth knowledge of contexts and established, working institutions. These can be key to more sustainable development approaches.

An anthropological approach proposes that it is possible to address the sustainability challenge: to create outcomes that are both multidimensional and lasting. To do so, developers must re-situate their work in the physical and social contexts that they aim to influence. One promising example that has emerged in recent years is *place-based development*.[22] Place-based development approaches vary in scope and effectiveness, but they aim to work *across sectors* while *embedding* change in existing local ways of working. Place-based approaches can link organisations in and beyond local places to tackle issues across sectors and silos. While requiring further investigation in practice, they may offer a way to meet the challenge of sustainability: how to create triple-bottom-line outcomes that last.

Summary: From challenges to opportunities?

This chapter has explored how ideas from anthropology can be used to unpack and reframe some of the current 'grand challenges' in twenty-first-century development work. Anthropological insights can explain a great deal about why mainstream approaches to doing development, fighting poverty, encouraging participation and pursuing sustainability often fail. Importantly, they can also suggest productive new ways to approach these challenges.

An anthropological reframing of development practice reveals previously invisible resources and strategies for change. Moving beyond project, program and policy frames, it considers how development initiatives play out on a broader development landscape. Anthropologists draw attention to the real-world contexts where poverty, participation and sustainability failures are experienced. At the same time, they reveal a range of actors, knowledges and institutions which can be vital allies and resources in any change-making process.

The message throughout this chapter has been that the 'grand challenges' of development work are serious but far from unsolvable. Development effectiveness, poverty eradication, authentic participation and sustainable change are at the heart of development aspirations globally and locally. Seen from within mainstream ways of 'doing development', however, they present intractable challenges. These challenges cannot be solved from within the frameworks that created them.

All development efforts – from high-level policy decisions to targeted local projects – hit the ground in real physical and social contexts. Ignoring these contexts, and the diverse actors who live and work in them, ignores the powerful engines of inertia and of change that surround us every day. It is akin to pulling a painted blind over a window to block the view. Development cannot be simply superimposed on a complex real-world landscape of people, organisations, agendas, worldviews and relationships. To be effective, to make a difference, to be inclusive and sustainable, it is necessary to engage with the social and cultural landscapes on which we work. This chapter proposes raising the metaphorical blind to re-engage with the development landscape – and by doing so, to find new ways of addressing old challenges.

The challenge of *development effectiveness* reminds us that development ignorance breeds failure, and failures are costly – often for those that can least afford them. Yet development ignorance is also eminently preventable. Most development failures come down to a basic misunderstanding of context: a belief that things are different than they are, or a failure to anticipate how contextual factors will influence outcomes. Policy makers currently rely on market signals or results-based measures to grapple with the challenge of ensuring effectiveness. Neither, however, has succeeded in making development more effective. An anthropological approach names the problem of development ignorance and seeks to overcome it by bringing local knowledge of context into decision-making processes. Because it matters who makes the decisions and what they know, development effectiveness is ultimately a question of development governance.

The challenge of *poverty* persists because 'poverty' is not a concrete thing that can be identified and eradicated; nor are the 'poor' a homogeneous group. Understanding poverty as a verb – a process – reframes anti-poverty work: from targeting 'the poor' as a problem to be solved, to targeting the relationships and actions that perpetuate poverty-as-process. An anthropological approach recognises that poverty happens in the relationships among development actors, in institutional contexts that place some people regularly and systematically at a disadvantage. More effective anti-poverty work asks: What are 'the poor' in particular contexts already doing to improve their situations – and what is blocking them from achieving their goals? What institutions and relationships need to change in order to stop producing disadvantage?

The challenge of *participation* reminds us that participants in development are actors in their own right, and that participation has costs, which are generally borne by those with the fewest resources. Further, the benefits for participants are often slight, with participants having little or no real say in decisions. 'Tacking on' participation to development business-as-usual makes it difficult for people to participate on their own terms. Shifting from *participation* to *recognition* reframes the role of developees: from participants in other people's processes, to actors with rights whose own knowledges and processes are recognised and respected. For professionals, this requires reflexivity: a willingness to question their own assumptions and be open to others' perspectives. Recognition is not only a more inclusive way of working, it opens the opportunity for innovative co-created solutions.

The challenge of *sustainability* highlights a shared commitment to triple-bottom-line development outcomes that are sustainable over time. Yet both aspects of sustainability are difficult to achieve within current development frameworks. Projects, programs and policies run to short-term time scales and generally deal with problems in silos rather than in context. An anthropological approach shows that it is possible to address the sustainability challenge by re-situating development action in physical and social contexts. Local actors and their institutions can illuminate on-the-ground connections among economic, social and environmental processes. They can also provide mechanisms to embed change over the long term. Place-based development approaches may provide a practical option for producing triple-bottom-line solutions that last.

Development practice in the twenty-first century faces a number of challenges that cut to the heart of our ability to be relevant, effective change makers. An anthropological approach suggests these challenges can also be seen as opportunities. In the end, no development professional or organisation, from a local council to the World Bank, needs to create development on their own. The world is full of development actors who can help. First, though, it is necessary to step reflexively outside our own frameworks and recognise the complex social and cultural landscapes on which we work. These landscapes have abundant and often surprising resources: local knowledges, diverse institutions, actors and relationships that can – if recognised – warn about the pitfalls, name the deeper challenges and drive new co-created solutions.

Notes

1 See e.g. Gardner and Lewis (2015, pp.64–65) for international aid work; they note that while there was growing demand for 'social development' expertise in the late 1980s, this waned in the 2000s with more macro-level management approaches and high-level international coordination. Increasingly, they note, 'aid has come to be seen as far more about managing things than about understanding people' (p.36).
2 See e.g. Mosse (2005).
3 Anthropologist Glynn Cochrane (2009, p.2) has described this as the Myth of Global Poverty: 'According to the Myth, the entire way of life of the poor would be eliminated using money, the weapon of mass development. . . . Poverty is defined as global, that is, the same everywhere.'
4 Mathur (2012).
5 Gardner and Lewis (2015, p.36).
6 For an analysis of the issues with current approaches to evidence-based development work in the international development context, see Eyben et al. (2015).
7 Cochrane (2009, p.18).
8 Gabriel (1991, p.2).
9 Cochrane (2009, p.3).
10 Green (2006, p.1124).
11 Gardner and Lewis (2015, p.45).
12 Reyna (1986, p.247).
13 Chapman (2014), quoted in Bauman et al. (2015, p.2).
14 Eversole (2016, p.102).
15 For instance, Venkatesan and Yarrow (2012, p.3) note that 'various post-development scholars have shown how development organizations define "problems" in ways that justify their own forms of "expertise" and thereby marginalize the insights and understandings of other groups of people.' Similarly, Odora Hoppers (2002, p.vii), writing on South African experiences, observes that: 'Sooner or later, the time will have to come to draw attention to the manner in which the exclusion of other traditions of knowledge by reductionist science is itself part of the problem that has led to myriad failed development initiatives all around the world.'
16 Cornwall (2010, p.4).
17 Eyben (2014, p.20).

18 Ibid.

19 Drinkwater (1992, p.170). He provides the caveat that, 'By definition, because we are not the other, such understandings can never be absolute.'

20 The widely cited definition of sustainable development from the Brundtland Report (United Nations 1987) is: 'Sustainable development is development that meets the needs of the present without compromising the ability of future generations to meet their own needs.'

21 Eade (2010, p.207).

22 For instance, The Prince's Charities Place Strategy in the UK (see e.g. https://prince ofwales.gov.uk/onlinereview2013/place-strategy/index.html), Collective Impact (http:// collaborationforimpact.com/collective-impact/) and other place-based approaches in Australia, and the Knowledge Partnering approach (Eversole 2015).

6

Conclusions: Using anthropology in development work

Social and cultural understanding is, more than ever before, vital for effective development practice. This book has argued that anthropology offers a unique perspective on development work – one that is currently missing as policy makers and practitioners grapple with the challenges of working with diverse communities and stakeholders, and meeting needs in vastly different contexts.

This book set out with an ambitious goal: to distil some of the key insights from the anthropology of development to inform practical development work. To do this, it has moved along the continuum from theory to practice: from core concepts and ideas from anthropologists who study development, to on-the-ground examples; then to frameworks and tools; and finally, to an exploration of how an anthropological approach can help development professionals respond to grand challenges toward the future.

Anthropologists have a long history of engaging with development work, both in theory and in practice. They have studied and directly practised development, and they have done so across both 'developed' and 'developing' country contexts. Many anthropologists have written compellingly about their insights and experiences. Yet it has been difficult for development professionals to make the connections between anthropological writing and development practice: to understand how ideas and insights from anthropology can be applied practically in development work.

This book has aimed to bridge this divide, by articulating an anthropological approach to development, and then showing how this approach can offer significant practical benefits for those who work to create positive change.

Chapter 1 began by identifying a set of key ideas that appear regularly when anthropologists write about development. These ideas about *contexts, actors, knowledges* and *institutions* have direct implications for development practice. They show why and how *culture* – shared ways of seeing, thinking and acting – matters for social change in general, and for development work in particular.

Chapter 1 explained that different people and organisations have different social roles and situations which affect their room to manoeuvre, and different cultural frames and logics that affect how they define and pursue positive change. In the end, the key insight is that development is always a social and cultural process.

Chapter 2 then illustrated how these ideas from the anthropology of development can be used to make sense of real-world development situations. Drawing on ethnographic case studies from around the world, Chapter 2 showed how both planned and unplanned processes of social and economic change can be understood with reference to the actions, knowledges and institutions of actors, their social positions and their relationships with one another. For development professionals, these case studies revealed both pitfalls and opportunities in practice, and suggested the need to reframe how we 'do' development to place people and relationships at the centre.

Chapter 3 took these key ideas from Chapters 1 and 2 and distilled them into a framework for development practice. The framework visualised what an anthropological approach to development looks like and how it differs from mainstream approaches to 'doing development'. An anthropological approach views all development initiatives on a broader development landscape. It shifts the focus from problems to contexts; from target groups to actors; and from carefully managed technical change to co-created solutions. The framework in Figure 3.2 can be a useful reference point to sum up in a visual way some of the key ideas in this book.

Chapter 4 then showed how this anthropological approach to development practice can be applied in day-to-day development work. The chapter is organised according to the standard processes of assessment, design, implementation and evaluation of development initiatives. Chapter 4 showed how development professionals can use a range of practice tools to embed attention to contexts, actors, knowledges and institutions at each stage of development work. This chapter also suggested some additional tools that can be used to encourage reflexivity – attention to the ideas and ways of working that guide our own practice, and openness to the ideas and ways of working of others – as well as when it can be useful to work with professional anthropologists.

Chapter 5 took a step back from day-to-day development practice to consider big-picture 'grand challenges' facing development work in the twenty-first century. Development effectiveness, poverty, participation and sustainability are significant challenges; Chapter 5 showed how an anthropological approach can help to develop concrete strategies to address them. Seeing development as a social and cultural process showed each of the 'grand challenges' in a new light. In particular, Chapter 5 highlighted the dangers of development ignorance and the power of reflexivity in development work. Reflexive practice can help reframe challenges of all kinds by moving beyond current ways of thinking and working to take on board new insights and practices. Reflexivity and reframing are incredibly useful skills in creative problem-solving.

To conclude, the following sections sum up three key 'take home' lessons. These are not new points; they re-state key messages from the book in a way that some readers may find useful and easy to remember.

Lesson one: Development abhors a vacuum

Always remember: Context matters.

There is a persisting myth that 'big-picture' development processes don't need to worry about context. International agreements, national policy statements and huge multi-site programs grapple with big, abstract problems (hunger, disease, climate change). They assume they can solve these problems in big, abstract ways.

Yet developers ignore context at their peril – and at the peril of those they aim to help. Decades of evidence tell us that one-size-fits-all solutions fail to fit many. Far-away decision makers often fail to understand the practicalities of local contexts, from physical terrain to social roles and institutions. Development ignorance regularly leads to ineffective, even damaging, outcomes.

Development never happens 'nowhere'. Every development initiative, no matter how big-picture, must hit the ground in real places, where it will affect real people, as well as the resources that are available, and who can access them; the rules that are in force, and who is advantaged or disadvantaged by them.

Furthermore, in real-world development landscapes, actors and institutions are already present. Their actions and interactions will influence if and how change actually happens, and who benefits or suffers as a result. Current social roles and institutions affect people's room to manoeuvre, and these often have deep historical roots. Understanding what is already there – from processes of poverty-creation to adaptive institutions that people 'on the margins' use to survive – is key to effective change.

The first key lesson from this book, therefore, is that development never takes place in a vacuum. Context always matters.

When anthropologists write about development, they frequently emphasise the *local* – local contexts, local knowledge, local institutions, local relationships. This is sometimes misinterpreted as a preference for 'small-scale' and 'grassroots' approaches – as opposed to the large-scale, multi-site initiatives that development organisations generally prefer. Yet anthropologists' *local* focus is not a question of scale, but of context.

Anthropologists call our attention to what development initiatives at any scale look like when seen up-close and from within – through the eyes of actors who, in their own eyes, are all local. Every big-picture national or global development problem manifests locally in particular social and physical contexts; these determine what the problem looks like, what causes it and what the options for change are. And every big-picture effort to solve problems also hits the ground in particular contexts, where actors and institutions are present that will shape the ultimate direction of change.

Global problems and global initiatives thus always play out in local contexts. Attacking problems without paying attention to context amounts to attacking an abstraction, a conceptual mirage. Anthropologists remind us that problems and solutions are different depending on where you are, who you are, and the resources, assets, options and room to manoeuvre available to you in your particular context. Development always takes place somewhere. It does not happen in a vacuum, but in dialogue with what is already there.

Lesson two: Development is always about people

Always remember: People create change.

Development is often portrayed as a technical endeavour: a process of defining problems and target groups, and designing and implementing solutions. Yet this technical approach disguises the social realities of how development actually works. A key lesson of this book is that development is a social and cultural process. It is always about people.

Development initiatives are designed, implemented, experienced, supported, reinterpreted and resisted by a wide range of people and their organisations. Their actions are both social and cultural, guided by their different ways of knowing, thinking and acting. Interactions among these actors create change or continuity. In practice development is not designed and implemented by experts; it is negotiated by people in a range of social roles.

Mainstream development practice still revolves around the roles of developers and their beneficiaries or target groups. This assumption of a developer–developee relationship, with the power relations it entails, is deeply embedded in the institutional apparatus of development work. Despite various efforts to implement more participatory approaches, this essential hierarchy of roles has remained intact. Developers tend to see themselves as the ones who are ultimately in charge of change.

An anthropological approach questions this developer-centric view, by reminding us that change can come from anywhere. Formal development initiatives are never simply implemented; they are negotiated among people – who may resist, reinterpret, sidetrack or reinvent them. All development initiatives affect and are affected by people, who are diverse, savvy and situated in social contexts.

Further, formal development initiatives are only one small part of the story. Social and economic change processes can be unplanned as well as planned. They can be intentional or unintentional. All development is instigated by people, championed by people, sidetracked or resisted by people. No one group of actors holds the reins of change alone.

All development, therefore, is social. 'Social' considerations are not a specific silo of development work. Nor are they a feel-good add-on to a technical process. Rather, paying attention to people and their organisations is key to understanding how development works.

Because all development is social, 'social' considerations like gender and ethnicity, or stakeholder engagement and participation, can never be simply add-ons to some technical main business of development work. Actors are at the centre of every process. Wealthy and poor, developers and developees, and – beyond these simplistic categories – people in a range of roles instigate and navigate change.

These people have differing amounts of resources and influence, different knowledges and logics about change, and they generally see things differently. Efforts to create positive change often reproduce the privileged ways of seeing and ways of working of more powerful actors. Yet the secrets to positive change are generally to be found elsewhere: in the invisible knowledges and experiences of *less* powerful actors.

Because development is about people, it is ultimately about relationships. In relationships among development actors, who is heard and who is silenced? What poverty-producing or prosperity-producing actions happen? Whose ways of working are respected or ignored? And how might different relationships create different results? These are the questions at the heart of development practice.

Lesson three: Reframing is the key to change

Always remember: Different ways of seeing create different solutions.

Because development is something that people do, it is possible to do it differently. Less powerful development actors have important knowledge, but what they know often remains invisible in development practice. Their ways of seeing and ways of working often fly under the radar. Too often, efforts to create change simply reproduce old power relationships in new clothes.

This book argues that *reflexivity* and *reframing* are processes that development professionals can use to do development differently.

Reflexivity is the process of becoming consciously aware of our own and others' ways of seeing the world. It involves naming our own frameworks and assumptions, and being open to other people's logics and views.

Reframing mobilises this awareness to move outside our habitual ways of seeing the world to take on board others' perspectives. Reframing development situations with attention to the knowledges, logics and ways of working of other development actors can open up unexpected ideas and solutions.

Currently, development organisations and professionals still frame themselves as the ultimate drivers of change. Developers define what needs to happen and how. Reflexivity encourages us to question our categories, our values and the ethics of our practice, and notice the value of what others know and do. Too often, development work is about trying to change people to look more like us, or more like we think they should be. The alternative is to recognise that there are many desired futures, and many ways to achieve positive change.

Recognising the value of what other development actors know and do enables development professionals to see development differently: from something that developers define and manage, to something that can be actively co-created with others. It makes it possible to reframe our own development problems and appreciate the nature of the development landscape that we are operating upon. It reveals the resources that are already there, and the realities of the room to manoeuvre that different actors in different situations have.

Reframing is a profound shift – it changes how we see ourselves, our work and the people and organisations we work with. It brings other actors, and their knowledges and practices, into active view, and reveals unexpected insights and resources, as well as new perspectives on old problems. In doing so, it creates spaces for new relationships to form and new ways of working to emerge.

Anthropology for development, toward the future

Anthropology is still a largely untapped resource for development work. Yet as old ways of 'doing development' struggle to respond to today's challenges, it is becoming clear that anthropology can play a key role in enabling change. Anthropology's insights about development as a social and cultural process provide a unique set of insights to facilitate and inform more effective and innovative development practice in the twenty-first century.

An anthropological approach to development practice reframes development: shifting the frame from problems, projects and programs, to people and relationships. In doing so, it brings diverse contexts and actors into view, and reveals previously invisible knowledges and institutions. An anthropological approach thus uncovers a range of untapped resources for change.

An anthropological approach encourages reflexivity, urging development professionals to consider how their own ways of seeing and ways of working influence how they define problems and implement solutions. It suggests that a willingness to step outside these frameworks can provide a promising path to development innovation.

In the twenty-first century, development is no longer something which the few attempt to design for the many. It is, as indeed it has always been, a process in which different actors interact to shape the direction of change. Anthropologists name this and describe it. They remind development professionals that development is, ultimately, something that people do. But development professionals, working within their own institutional and organisational constraints, must decide how they will engage with these insights.

Into the future, people and organisations will continue to interact on the development landscape, negotiating social and economic change as best they can in the contexts where they live and work. Whether or not developers choose to acknowledge or understand these actors, their interactions and relationships will define who benefits from change and who does not. They will define whether poverty-as-process is accelerated or eradicated. The question then becomes, how will the actions of developers – their projects, programs and policies – interact with the actions of other actors? Will the results slow or speed up poverty? Will they reinforce, or reinvent, the status quo?

The practical opportunity for development professionals is to recognise and work with this social energy. An anthropological view reveals a development landscape where there are already many allies for change. Developers don't have to solve the world's problems from afar, or alone. In the end, the world is full of people with deep knowledge of their own contexts, who are passionate about the changes that matter to them and whose insights have the potential to stimulate significant innovation. Opportunities to challenge the relationships that perpetuate poverty are abundant, and even small shifts can be powerful – simply by recognising what others know, how they work and what they can teach us about change.

Further reading

Anthropologists have written about development in a wide range of contexts. Readers interested in learning more about anthropological approaches to development are encouraged to explore this rich literature, and to seek out studies that resonate with their own specific areas of geographic and/or topical interest.

Research in the anthropology of development has often been published in scholarly journals, but there is no single journal dedicated to anthropology of development. Finding relevant articles thus generally requires a trawl through databases. Journals in development studies occasionally publish anthropological work on development, as do some anthropology journals, as well as journals in fields such as public policy and social research. The applied anthropology journal *Human Organization* is an excellent resource for practically focused anthropological analyses and regularly publishes development-related work.

Various anthropology of development studies are published as sole-authored or edited books. Some are specifically dedicated to development-related questions – 'development' is front and centre in the title – while others deal with development as a part of larger studies. Works in economic and political anthropology in particular are often relevant to the anthropology of development; they explain dynamics such as the growth of markets or the changing nature of production in particular contexts, or analyse how particular institutions create and perpetuate status and disadvantage.

The studies cited in this book include an eclectic mix of 'classics' and more recent contributions. A number of the cases are from the 1980s and 1990s, which reflects the abundance of rich ethnographic literature on development in those decades. Yet good, ethnographically rich anthropology of development continues to be published right up until the present day, providing a fantastic resource for understanding development processes and development practices, seen anthropologically.

The following annotated reading list provides some guidance for further reading; these texts are both insightful and generally accessible, with rich case material on development actors, knowledges and institutions. The readings have been organised into clusters for ease of reference.

Development case studies – useful collections

Bush Base, Forest Farm: Culture, Environment and Development, edited by Elisabeth Croll and David Parkin. Routledge (1992).

An Anthropological Critique of Development: The Growth of Ignorance, edited by Mark Hobart. Routledge (1993).

Discourses of Development: Anthropological Perspectives, edited by R.D. Grillo and R.L. Stirrat. Bloomsbury (1997).

Anthropological Perspectives on Local Development: Knowledge and Sentiments in Conflict, edited by Simone Abram and Jacqueline Waldren. Routledge (1998).

Participating in Development: Approaches to Indigenous Knowledge, edited by Paul Sillitoe, Alan Bicker and Johan Pottier. Routledge (2002).

Negotiating Local Knowledge: Power and Identity in Development, edited by Johan Pottier, Alan Bicker and Paul Sillitoe. Pluto Press (2003).

Investigating Local Knowledge: New Directions, New Approaches, edited by Alan Bicker, Paul Sillitoe and Johan Pottier. Ashgate (2004).

Several excellent anthropological collections were published in the 1990s and 2000s which looked up-close at development projects and processes playing out in particular contexts, and analysed the actors, institutions, knowledges and processes involved. All of these collections have broad geographic and topical range; many of the individual chapters were based on in-depth ethnographic research in different parts of the world.

The collection by Croll and Parkin has a particular focus on the environmental aspects of local practices and the interplay of local resource management practices with external development initiatives. The collections by Hobart, Sillitoe, Pottier and Bicker emphasise local and indigenous knowledges with particular reference to the intersection of different kinds of knowledge in development processes. The collection by Grillo and Stirrat analyses development discourse with reference to on-the-ground ethnographies of development practice, and the collection by Abram and Waldren similarly explores some of the tensions between local and external perspectives on development processes in the context of industrialised countries.

Ethnographies of local economic development

Peddlers and Princes: Social Development and Economic Change in Two Indonesian Towns, by Clifford Geertz. University of Chicago Press (1963).

Raising Cane: The Political Economy of Sugar in Western India, by Donald Attwood. Westview Press (1992).

Smallholders, Householders, Farm Families and the Ecology of Intensive, Sustainable Agriculture, by Robert McC. Netting. Stanford University Press (1993).

Making a Market: The Institutional Transformation of an African Society, by Jean Ensminger. Cambridge University Press (1996).

Between Field and Cooking Pot: The Political Economy of Marketwomen in Peru, by Florence Babb. University of Texas Press (1998).

Weaving a Future: Tourism, Cloth, and Culture on an Andean Island, by Elayne Zorn. University of Iowa Press (2004).

A number of anthropological studies look at questions of economic change in particular social contexts. Anthropologists seat their economic analyses in a deep understanding of the social and cultural embeddedness of economic action. These studies provide an up-close, in-depth ethnographic view of how economic development happens, with attention to politics, institutions and the actions of local actors. A key message is often about the savvy ways that local actors and their institutions navigate economic, social and environmental challenges.

Excellent examples of this genre of work include: Clifford Geertz's analysis of local economic actors in two regions of Indonesia; Donald Attwood's study of the evolution of peasant-run and state-run sugar-cane cooperatives in Western India; Robert Netting's analysis of production strategies employed by smallholder farmers across different local contexts; Jean Ensminger's study of the emergence of formal markets among pastoralists in Kenya; Florence Babb's up-close look at the strategies employed by market vendors in the Peruvian informal economy; and Elayne Zorn's description of the development of community-run tourism initiatives in Ecuador.

Unpacking the idea of development

The Anti-Politics Machine: 'Development,' Depoliticization, and Bureaucratic Power in Lesotho, by James Ferguson. University of Minnesota Press (1994).

Encountering Development: The Making and Unmaking of the Third World, by Arturo Escobar. Princeton University Press (1995, 2012).

'Representing Poverty and Attacking Representations: Perspectives on Poverty from Social Anthropology', by Maia Green. *Journal of Development Studies* 42(7): 1108–1129 (2006).

Festival Elephants and the Myth of Global Poverty, by Glynn Cochrane. Pearson Education (2009).

Deconstructing Development Discourse: Buzzwords and Fuzzwords, edited by Andrea Cornwall and Deborah Eade. Practical Action Publishing (2010).

'Paradigm of "Better Life": "Development" among the Khumi in the Chittagong Hill Tracts', by Nasir Uddin. *Asian Ethnicity* 15(1): 62–77 (2013).

'"They Are Not Understanding Sustainability": Contested Sustainability Narratives at a Northern Malawian Development Interface', by Thomas McNamara. *Human Organization* 76(2): 121–130 (2017).

A number of anthropological studies provide a critical analysis of mainstream ways of thinking about and talking about development. In doing so, they serve to 'name up' often unconscious assumptions and values that frame development practice in certain ways. Many of these studies examine development *discourse*: that is, how policy makers, developers and others think about and talk about development. Discourse matters because it reveals a great deal about the ideas that guide practice. This literature shows how developers often frame their work in certain ways, and these framings influence what is seen, and what is left out.

In James Ferguson's analysis of a development project in Lesotho, he shows how development was framed to leave out key aspects of politics and history, and focus narrowly on technical problems. Arturo Escobar's more wide-ranging critique argues that 'development' created the idea of underdevelopment and imposed it on post-colonial countries. Maia Green's article from the *Journal of Development Studies* explores the idea of poverty, both as a label applied to certain categories of people, and as a category of development thinking; while Glynn Cochrane's book unpacks the discourses of 'global poverty' and what they mean in practice. The edited collection by Andrea Cornwall and Deborah Eade provides an accessible introduction to the analysis of development discourse by providing a critical perspective on key development terms – which they term 'Developmentspeak'. Finally, the article by Nasir Uddin contrasts global development discourse with the very different local perspectives of the Khumi people of Bangladesh, and the article by Thomas McNamara shows how rural people and NGO fieldworkers in Malawi had very different ideas about the nature of sustainable development.

Anthropology of / for practice

Development Economics on Trial: The Anthropological Case for a Prosecution, by Polly Hill. Cambridge University Press (1986).

'Dams, Cows, and Vulnerable People: Anthropological Contributions to Sustainable Development', by Michael M. Horowitz. *The Pakistan Development Review* 34(4): 481–508 (1995).

Cultivating Development: An Ethnography of Aid Policy and Practice, by David Mosse. Pluto Press (2005).

Adventures in Aidland: The Anthropology of Professionals in International Development, edited by David Mosse. Berghahn Books (2011).

'Performing Partnership: Invited Participation and Older People's Forums', by Elizabeth Harrison. *Human Organization* 71(2): 157–166 (2012).

'The Pragmatics of Ethnography and Intervention: Street Children in Mexico', by Norman Long. *European Journal of Development Research* 25(3): 356–366 (2013).

International Aid and the Making of a Better World: Reflexive Practice, by Rosalind Eyben. Routledge (2014).

Regional Development in Australia: Being Regional, by Robyn Eversole. Routledge (2016).

A number of anthropological analyses of development projects, programs and policies are available; these studies aim to explicitly use anthropological approaches to understand the nature and limitations of current ways of working in development practice. These studies explore the practical insights that an anthropological approach can provide to inform professional development work.

Polly Hill's *Development Economics on Trial* is a classic in this space; she demonstrates the gaps in economic analyses of rural development processes and argues for the need for anthropological insights to inform development efforts. Similarly, the work of Michael Horowitz and his colleagues at the US-based Institute for Development Anthropology brought insights from anthropology to World Bank programs in Africa. David Mosse's ethnography of aid work in India provides on-the-ground insights into the social processes of implementing development policies, highlighting the limitations of 'good policy' for achieving development outcomes. His later edited collection brings together a number of anthropologists involved in development practice to explore how development professionals 'do' international aid.

Rosalind Eyben is an anthropologist and development professional; in *International Aid and the Making of a Better World* her engaging reflections on her decades of international development work in numerous contexts build a compelling case for the importance of reflexive practice. In the UK context, Elizabeth Harrison explores the perspectives of older people on the limitations of local council efforts at participation, while in Mexico, Norman Long demonstrates the value of ethnography in understanding the perspectives of street children on the programs designed to help them. Finally, *Regional Development in Australia: Being Regional* is a consciously anthropological analysis of the nature and limits of contemporary Australian regional development policy.

References

Abram, S. (1998) 'Introduction: Anthropological Perspectives on Local Development', in *Anthropological Perspectives on Local Development: Knowledge and Sentiments in Conflict*, eds. S. Abram and J. Waldren. London: Routledge.

Abram, S. and J. Waldren (eds.) (1998) *Anthropological Perspectives on Local Development: Knowledge and Sentiments in Conflict*. London: Routledge.

Acheson, J.M. (1988) *The Lobster Gangs of Maine*. Lebanon, NH: University Press of New England.

Appadurai, A. (2004) 'The Capacity to Aspire: Culture and the Terms of Recognition', in *Culture and Public Action*, eds. V. Rao and M. Walton. Stanford, CA: Stanford University Press.

Argyris, C. and D. Schön (1974) *Theory in Practice: Increasing Professional Effectiveness*. San Francisco, CA: Jossey-Bass.

Argyris, C. and D. Schön (1978) *Organizational Learning: A Theory of Action Perspective*. Reading, MA: Addison Wesley.

Attwood, D.W. (1992) *Raising Cane: The Political Economy of Sugar in Western India*. Boulder, CO: Westview Press.

Attwood, D.W. (1997) 'The Invisible Peasant', in *Economic Analysis Beyond the Local System*, ed. R. Blanton *et al.* Society for Economic Anthropology Monograph Series. Lanham, MD: University Press of America.

Ayers, A. (1995) 'Indigenous Soil and Water Conservation in Djenné, Mali', in *The Cultural Dimension of Development: Indigenous Knowledge Systems*, eds. D.M. Warren, L.J. Slikkerveer and D. Brokensha. London: Intermediate Technology Publications.

Babb, F. (1998) *Between Field and Cooking Pot: The Political Economy of Marketwomen in Peru*. Austin, TX: University of Texas Press.

Bauman, T., D. Smith, R. Quiggin, C. Keller and L. Drieberg (2015) 'Building Aboriginal and Torres Strait Islander Governance: Report of a Survey and Forum to Map Current and Future Research and Practical Resource Needs.' Report (May). Canberra: Australian Institute of Aboriginal and Torres Strait Islander Studies (AIATSIS).

Bicker, A., P. Sillitoe and J. Pottier (eds.) (2004) *Investigating Local Knowledge: New Directions, New Approaches*. Aldershot: Ashgate.

Bodley, J.H. (1990) *Victims of Progress* (Third Edition). Mountain View, CA: Mayfield Publishing.

Boissevain, J. and N. Theuma (1998) 'Contested Space: Planners, Tourists, Developers and Environmentalists in Malta', in *Anthropological Perspectives on Local Development: Knowledge and Sentiments in Conflict*, eds. S. Abram and J. Waldren. London: Routledge.

Brokensha, D., D. Warren and O. Werner (eds.) (1980) *Indigenous Knowledge Systems and Development*. Lanham, MD: University Press of America.

Brown, J., D. Isaacs and the World Café Community (2005) *The World Café: Shaping Our Future Through Conversations that Matter*. San Francisco, CA: Berrett-Koehler.

Burns, D. and S. Worsley (2015) *Navigating Complexity in International Development: Facilitating Sustainable Change at Scale*. Rugby, UK: Practical Action Publishing.

Cernea, M.M. (1986) 'Foreword', in *Anthropology and Rural Development in West Africa*, eds. M.M. Horowitz and T.M. Painter. Boulder, CO: Westview Press.

Chambers, R. (1983) *Rural Development: Putting the Last First*. Harlow, UK: Pearson Education.

Chambers, R. (1994) 'The Origins and Practice of Participatory Rural Appraisal', *World Development* 22(7): 953–969.

Chambers, R. (2002) *Participatory Workshops: A Sourcebook of 21 Sets of Ideas and Activities*. London: Earthscan.

Chambers, R., A. Pacey and L. Thrupp (eds.) (1989) *Farmer First: Farmer Innovation and Agricultural Research*. London: Intermediate Technology Publications.

Cochrane, G. (1979) *The Cultural Appraisal of Development Projects*. New York and London: Praeger.

Cochrane, G. (2009) *Festival Elephants and the Myth of Global Poverty*. Boston, MA: Pearson Education.

Cornwall, A. (2008) 'Unpacking "Participation": Models, Meanings and Practices', *Community Development Journal* 43(3): 269–283.

Cornwall, A. (2010) 'Introductory Overview – Buzzwords and Fuzzwords: Deconstructing Development Discourse', in *Deconstructing Development Discourse: Buzzwords and Fuzzwords,* eds. A. Cornwall and D. Eade. Rugby, UK: Practical Action Publishing.

Cornwall, A. and D. Eade (eds.) (2010) *Deconstructing Development Discourse: Buzzwords and Fuzzwords*. Rugby, UK: Practical Action Publishing.

Crewe, E. (1997) 'The Silent Traditions of Developing Cooks', in *Discourses of Development: Anthropological Perspectives*, eds. R.D. Grillo and R.L. Stirrat. Oxford: Bloomsbury.

Crewe, E. (2014) 'Doing Development Differently: Rituals of Hope and Despair in an INGO', *Development in Practice* 24(1): 91–104.

Drinkwater, M. (1992) 'Cows Eat Grass Don't They? Evaluating Conflict over Pastoral Management in Zimbabwe', in *Bush Base: Forest Farm, Culture, Environment and Development*, eds. E. Croll and D. Parkin. New York: Routledge.

Eade, D. (2010) 'Capacity Building: Who Builds Whose Capacity?', in *Deconstructing Development Discourse: Buzzwords and Fuzzwords,* eds. A. Cornwall and D. Eade. Rugby, UK: Practical Action Publishing.

Escobar, A. (1995, 2012) *Encountering Development: The Making and Unmaking of the Third World*. Princeton, NJ: Princeton University Press.

Esman, M. and N. Uphoff (1984) *Local Organizations: Intermediaries in Rural Development*. Ithaca, NY: Cornell University Press.

Eversole, R. (2012) 'Remaking Participation: Challenges for Community Development Practice', *Community Development Journal*, 47(1): 29–41.

Eversole, R. (2015) *Knowledge Partnering for Community Development*. New York: Routledge.

Eversole, R. (2016) *Regional Development in Australia: Being Regional*. New York: Routledge.

Eversole, R., J. McNeish and A. Cimadamore (eds.) (2005) *Indigenous Peoples and Poverty in International Perspective*. London: Zed Books.

Eyben, R. (ed.) (2006) *Relationships for Aid*. London: Earthscan.

Eyben, R. (2014) *International Aid and the Making of a Better World: Reflexive Practice*. New York: Routledge.

Eyben, R., I. Guijt, C. Roche and C. Shutt (eds.) (2015) *The Politics of Evidence and Results in International Development: Playing the Game to Change the Rules?* Rugby, UK: Practical Action Publishing.

Eyben, R., C. Harris and J. Pettit (2006) 'Introduction: Exploring Power for Change', *IDS Bulletin* 37(6): 1–10.

Eyben, R. and R. León (2006) 'Whose Aid? The Case of the Bolivian Elections Project', in *The Aid Effect*, eds. D. Lewis and D. Mosse. London: Pluto Press.

Fairhead, J. (1993) 'Representing Knowledge: The "New Farmer" in Research Fashions', in *Practicing Development: Social Science Perspectives*, ed. J. Pottier. New York: Routledge.

Fairhead, J. and M. Leach (1997) 'Webs of Power and the Construction of Environmental Policy Problems: Forest Loss in Guinea', in *Discourses of Development: Anthropological Perspectives*, eds. R.D. Grillo and R.L. Stirrat. Oxford: Bloomsbury.

Ferguson, J. (1994) *The Anti-Politics Machine: 'Development,' Depoliticization, and Bureaucratic Power in Lesotho*. Minneapolis: University of Minnesota Press.

Fujisaka, S. (1995) 'Taking Farmers' Knowledge and Technology Seriously: Upland Rice Production in the Philippines', in *The Cultural Dimension of Development: Indigenous Knowledge Systems*, eds. D.M. Warren, L.J. Slikkerveer and D. Brokensha. London: Intermediate Technology Publications.

Gabriel, T. (1991) *The Human Factor in Rural Development*. London: Belhaven.

Gamser, M. and H. Appleton (1995) 'Tinker, Tiller, Technical Change: Peoples' Technology and Innovation Off the Farm', in *The Cultural Dimension of Development: Indigenous Knowledge Systems*, eds. D.M. Warren, L.J. Slikkerveer and D. Brokensha. London: Intermediate Technology Publications.

Gardner, K. and D. Lewis (2015) *Anthropology and Development: Challenges for the Twenty-First Century*. London: Pluto Press.

Geertz, C. (1963) *Peddlers and Princes: Social Development and Economic Change in Two Indonesian Towns*. Chicago, IL: University of Chicago Press.

Green, M. (2006) 'Representing Poverty and Attacking Representations: Perspectives on Poverty from Social Anthropology', *Journal of Development Studies* 42(7): 1108–1129.

Green, M. (2012) 'Framing and Escaping, Contrasting Aspects of Knowledge Work in International Development and Anthropology', in *Differentiating Development: Beyond an Anthropology of Critique*, eds. S. Venkatesan and T. Yarrow. New York: Berghahn Books.

Grillo, R.D. and R.L. Stirrat (eds.) (1997) *Discourses of Development: Anthropological Perspectives*. Oxford: Bloomsbury.

Haddad, L., N. Hossain, J.A. McGregor and L. Mehta (2011) 'Introduction: Time to Reimagine Development?' *IDS Bulletin* 42(September): 1–12.

Hamilton, S. (1998) *The Two-Headed Household: Gender and Rural Development in the Ecuadorean Andes*. Pittsburgh, PA: The University of Pittsburgh Press.

Harrison, E. (2012) 'Performing Partnership: Invited Participation and Older People's Forums', *Human Organization* 71(2): 157–166.

Harrison, E. (2013) 'Beyond the Looking Glass? "Aidland" Reconsidered', *Critique of Anthropology* 33(3): 263–279.

Harrison, E. (2015) 'Anthropology and Impact Evaluation: A Critical Commentary', *Journal of Development Effectiveness* 7(2): 146–159.

Hart, G. (2001) 'Development Critiques in the 1990s: Cul-de-sac and Promising Paths', *Progress in Human Geography* 25(4): 649–658.

Hausner, S. (2006) 'Anthropology in Development: Notes from an Ethnographic Perspective', *India Review* 5(3–4): 318–342.

Hill, P. (1986) *Development Economics on Trial: The Anthropological Case for a Prosecution*. Cambridge: Cambridge University Press.

Hobart, M. (ed.) (1993) *An Anthropological Critique of Development: The Growth of Ignorance*. London: Routledge.

Hoben, A. (1982) 'Anthropologists and Development', *Annual Review of Anthropology* 11: 349–375.

Hopper, P. (2012) *Understanding Development*. Cambridge: Polity Press.

Horowitz, M.M. (1995) 'Dams, Cows, and Vulnerable People: Anthropological Contributions to Sustainable Development', *The Pakistan Development Review* 34(4): 481–508.

Irvine, R., R. Chambers and R. Eyben (2004) 'Learning from Poor People's Experience: Immersions'. *Lessons for Change in Policy and Organisations* Series. Brighton, UK: Institute of Development Studies.

Kahakalau, K. (2016) 'Pedagogy of Aloha - Trusting in our Native Traditions', Japanaka errol West Annual Lecture presented at the University of Tasmania, July. Available at: https://livestream.com/UniversityofTasmania/events/5577678, viewed 13 December 2016.

Larsen, A.K. (1998) 'Discourses of Development in Malaysia', in *Anthropological Perspectives on Local Development: Knowledge and Sentiments in Conflict*, eds. S. Abram and J. Waldren. London: Routledge.

Leach, M. (1992) 'Women's Crops in Women's Spaces, Gender Relations in Mende Rice Farming', in *Bush Base: Forest Farm, Culture, Environment and Development*, eds. E. Croll and D. Parkin. London: Routledge.

Leach, M. (2015) 'The Ebola Moment: Mobilising Engaged Anthropology for Global Health', Plenary Presentation at *MAGic2015: Anthropology and Global Health: Interrogating Theory, Policy and Practice*, European Association of Social Anthropologists, University of Sussex, September.

Lee, E. (2016) 'Protected Areas, Country and Value: The Nature–Culture Tyranny of the IUCN's Protected Area Guidelines for Indigenous Australians', *Antipode* 48(2): 355–374.

Lewis, D. (2009) 'International Development and the "Perpetual Present": Anthropological Approaches to the Re-Historicization of Policy', *European Journal of Development Research* 21(1): 32–46.

Long, N. (2001) *Development Sociology: Actor Perspectives*. New York: Routledge.

Long, N. (2013) 'The Pragmatics of Ethnography and Intervention: Street Children in Mexico', *European Journal of Development Research* 25(3): 356–366.

McNamara, T. (2017) '"They Are Not Understanding Sustainability": Contested Sustainability Narratives at a Northern Malawian Development Interface', *Human Organization* 76(2): 121–130.

Mairal Buil, G. and J.A. Bergua (1998) 'From Economism to Culturalism: The Social and Cultural Construction of Risk in the River Esera (Spain)', in *Anthropological Perspectives on Local Development: Knowledge and Sentiments in Conflict*, eds. S. Abram and J. Waldren. London: Routledge.

Mathur, N. (2012) 'Effecting Development and the Effects of Development: Bureaucratic Knowledges of Development in an Indian District', in *Differentiating Development: Beyond an Anthropology of Critique*, eds. S. Venkatesan and T. Yarrow. New York: Berghahn Books.

Mizanuddin, M. (ed.) (2013) *Ideas and Practices in Development: Anthropological Perspective*. Rajshahi, Bangladesh: Department of Anthropology, Rajshahi University.

Mosse, D. (1997) 'The Ideology and Politics of Community Participation', in *Discourses of Development: Anthropological Perspectives*, eds. R.D. Grillo and R.L. Stirrat. Oxford: Bloomsbury.

Mosse, D. (2005) *Cultivating Development: An Ethnography of Aid Policy and Practice*. London: Pluto Press.

Munson, H. Jr. (1990) 'Slash-and-Burn Cultivation, Charcoal Making, and Emigration from the Highlands of Northwest Morocco', in *Anthropology and Development in North Africa and the Middle East*, eds. M. Salem-Murdock, M.M. Horowitz and M. Sella. Boulder, CO and Birmingham, NY: Institute for Development Anthropology and Westview Press.

Netting, R.McC. (1991) 'Practicing the Anthropology of Development: Comments on "Anthropology and Foreign Assistance"', *Studies in Third World Societies* 44: 147–152.

Niamir, M. (1995) 'Indigenous Systems of Natural Resource Management among Pastoralists of Arid and Semi-arid Africa', in *The Cultural Dimension of Development: Indigenous Knowledge Systems*, eds. D.M. Warren, L.J. Slikkerveer and D. Brokensha. London: Intermediate Technology Publications.

North, D. (1990) *Institutions, Institutional Change and Economic Performance*. Cambridge: Cambridge University Press.

Odora Hoppers, C.A. (ed.) (2002) *Indigenous Knowledge and the Integration of Knowledge Systems*. Claremont, South Africa: New Africa Books.

Olivier de Sardan, J.P. (2005) *Anthropology and Development: Understanding Contemporary Social Change*. London: Zed Books.

Olwig, M. (2013) 'Beyond Translation: Reconceptualizing the Role of Local Practitioners and the Development "Interface"', *European Journal of Development Research* 25(3): 428–444.

Ortner, S.B. (1984) 'Theory in Anthropology since the Sixties', *Comparative Studies in Society and History* 26(1): 126–166.

Peattie, L.R. (1991) 'Planning and the Image of the City', *Places* 7(2): 35–39.

Pottier, J. (ed.) (1993) *Practising Development: Social Science Perspectives*. London and New York: Routledge.

Pottier, J. (1997) 'Towards an Ethnography of Participatory Appraisal and Research', in *Discourses of Development: Anthropological Perspectives*, eds. R.D. Grillo and R.L. Stirrat. Oxford: Bloomsbury.

Pottier, J., A. Bicker and P. Sillitoe (eds.) (2003) *Negotiating Local Knowledge: Power and Identity in Development*. London: Pluto Press.

Reyna, S.P. (1986) 'Donor Investment Preference, Class Formation, and Existential Development: Articulation of Production Relations in Burkina Faso', in *Anthropology and Rural Development in West Africa*, eds. M.M. Horowitz and T.M. Painter. Boulder, CO: Westview Press.

Rhoades, R. and A. Bebbington (1995) 'Farmers Who Experiment: An Untapped Resource for Agricultural Research and Development', in *The Cultural Dimension of Development: Indigenous Knowledge Systems*, eds. D.M. Warren, L.J. Slikkerveer and D. Brokensha. London: Intermediate Technology Publications.

Richards, P. (1985) *Indigenous Agricultural Revolution: Ecology and Food Production in West Africa*. London: Hutchinson.

Richards, P. (1993) 'Cultivation: Knowledge or Performance?', in *An Anthropological Critique of Development: The Growth of Ignorance*, ed. M. Hobart. London and New York: Routledge.

Rusten, E.P. and M.A. Gold (1995) 'Indigenous Knowledge Systems and Agro-Forestry Projects in the Central Hills of Nepal', in *The Cultural Dimension of Development: Indigenous Knowledge Systems*, eds. D.M. Warren, L.J. Slikkerveer and D. Brokensha. London: Intermediate Technology Publications.

Schön, D.A. (1984) *The Reflective Practitioner: How Professionals Think In Action*. New York: Basic Books.

Scoones, I. and J. Thompson (eds.) (1994) *Beyond Farmer First: Rural People's Knowledge, Agricultural Research, and Extension Practice*. London: Intermediate Technology Publications.

Scoones, I., J. Thompson and R. Chambers (eds.) (2009) *Farmer First Revisited: Innovation for Agricultural Research and Development*. Rugby, UK: Practical Action Publishing.

Sharland, R.W. (1995) 'Using Indigenous Knowledge in a Subsistence Society of Sudan', in *The Cultural Dimension of Development: Indigenous Knowledge Systems*, eds. D.M. Warren, L.J. Slikkerveer and D. Brokensha. London: Intermediate Technology Publications.

Shaw, R. (1992) '"Nature", "Culture" and Disasters: Floods and Gender in Bangladesh', in *Bush Base: Forest Farm, Culture, Environment and Development*, eds. E. Croll and D. Parkin. New York: Routledge.

Sillitoe, P. (2015) 'The Dialogue between Indigenous Studies and Engaged Anthropology: Some First Impressions', in *Indigenous Studies and Engaged Anthropology: The Collaborative Moment*, ed. P. Sillitoe. Farnham, UK: Ashgate.

Sillitoe, P., A. Bicker and J. Pottier (eds.) (2002) *Participating in Development: Approaches to Indigenous Knowledge*. ASA Monographs 39. London: Routledge.

Simonelli, J. and D. Earle (2003) 'Disencumbering Development: Alleviating Poverty through Autonomy in Chiapas', in *Here to Help: NGOs Combating Poverty in Latin America*, ed. R. Eversole. Armonk, NY: M.E. Sharpe.

Skinner, J. (2003) 'Anti-Social "Social Development"? Governmentality, Indigenousness and the DFID Approach on Montserrat', in *Negotiating Local Knowledge: Power and Identity in Development*, eds. J. Pottier, A. Bicker and P. Sillitoe. London: Pluto Press.

Smith, B.R. (2003) '"All Has Been Washed Away Now": Tradition, Change and Indigenous Knowledge in a Queensland Aboriginal Land Claim', in *Negotiating Local Knowledge: Power and Identity in Development*, eds. J. Pottier, A. Bicker and P. Sillitoe. London: Pluto Press.

Smyth, I. (2010) 'Talking of Gender: Words and Meanings in Development Organisations', in *Deconstructing Development Discourse: Buzzwords and Fuzzwords*, eds. A. Cornwall and D. Eade. Rugby, UK: Practical Action Publishing.

Soliz Tito, L. (2011) 'Apuesta Audaz por Modelos de Desarrollo Emergentes y Actores Marginados', Paper presented at the International Seminar, 'Modelos de Desarrollo, Desarrollo Rural y Economía Campesina Indígena', La Paz, April.

Standing, H. (2013) 'Introduction', in *Ideas and Practices in Development: Anthropological Perspective,* ed. M. Mizanuddin. Rajshahi, Bangladesh: Department of Anthropology, Rajshahi University.

Standing, H., K. Hawkins, E. Mills, S. Theobald and C.C. Undie (2011) 'Introduction: Contextualising "Rights" in Sexual and Reproductive Health', *BMC International Health and Human Rights* 11 (Supplement 3). Available at: http://bmcinthealthhumrights.biomedcentral.com/articles/10.1186/1472-698X-11-S3-S1, viewed 13 December 2016.

Talle, A. (1998) 'Sex for Leisure: Modernity among Female Bar Workers in Tanzania', in *Anthropological Perspectives on Local Development: Knowledge and Sentiments in Conflict,* eds. S. Abram and J. Waldren. London: Routledge.

Tendler, J. (1975) *Inside Foreign Aid.* Baltimore, MD: The Johns Hopkins University Press.

Thurston, H.D. and J.M. Parker (1995) 'Raised Beds and Plant Disease Management', in *The Cultural Dimension of Development: Indigenous Knowledge Systems,* eds. D.M. Warren, L.J. Slikkerveer and D. Brokensha. London: Intermediate Technology Publications.

Uddin, N. (2013) 'Paradigm of "Better Life": "Development" among the Khumi in the Chittagong Hill Tracts', *Asian Ethnicity* 15(1): 62–77.

United Nations (UN) (1987) *Our Common Future: Report of the World Commission for Environment and Development* ('Brundtland Report'). New York: United Nations.

United Nations (UN) (2015) *Transforming our World: The 2030 Agenda for Sustainable Development.* New York: United Nations. Available at: https://sustainabledevelopment.un.org/post2015/transformingourworld/publication, viewed 8 December 2016.

Venkatesan, S. and T. Yarrow (eds.) (2012) *Differentiating Development: Beyond an Anthropology of Critique.* New York: Berghahn Books.

Warren, D.M., L.J. Slikkerveer and D. Brokensha (eds.) (1995) *The Cultural Dimension of Development: Indigenous Knowledge Systems.* London: Intermediate Technology Publications.

Wolf, E. (1982) *Europe and the People without History.* Berkeley, CA: University of California Press.

Zorn, E. (2004) *Weaving a Future: Tourism, Cloth, and Culture on an Andean Island.* Iowa City: University of Iowa Press.

Index

Page numbers in **bold** indicate a Box or Table, *italics* an illustration and n, an endnote

Taylor & Francis eBooks

Helping you to choose the right eBooks for your Library

Add Routledge titles to your library's digital collection today. Taylor and Francis ebooks contains over 50,000 titles in the Humanities, Social Sciences, Behavioural Sciences, Built Environment and Law.

Choose from a range of subject packages or create your own!

Benefits for you

» Free MARC records
» COUNTER-compliant usage statistics
» Flexible purchase and pricing options
» All titles DRM-free.

Benefits for your user

» Off-site, anytime access via Athens or referring URL
» Print or copy pages or chapters
» Full content search
» Bookmark, highlight and annotate text
» Access to thousands of pages of quality research at the click of a button.

REQUEST YOUR FREE INSTITUTIONAL TRIAL TODAY

Free Trials Available
We offer free trials to qualifying academic, corporate and government customers.

eCollections – Choose from over 30 subject eCollections, including:

Archaeology	Language Learning
Architecture	Law
Asian Studies	Literature
Business & Management	Media & Communication
Classical Studies	Middle East Studies
Construction	Music
Creative & Media Arts	Philosophy
Criminology & Criminal Justice	Planning
Economics	Politics
Education	Psychology & Mental Health
Energy	Religion
Engineering	Security
English Language & Linguistics	Social Work
Environment & Sustainability	Sociology
Geography	Sport
Health Studies	Theatre & Performance
History	Tourism, Hospitality & Events

For more information, pricing enquiries or to order a free trial, please contact your local sales team: www.tandfebooks.com/page/sales

Routledge
Taylor & Francis Group

The home of
Routledge books

www.tandfebooks.com